# THE TECHNICAL
# INSTRUCTOR'S HANDBOOK
## *From Techie to Teacher*

# Books and Training Products From QED

## DATABASE

Data Analysis: The Key to Data Base Design
Diagnostic Techniques for IMS Data Bases
The Data Dictionary: Concepts and Uses
DB2: The Complete Guide to
  Implementation and Use
Logical Data Base Design
DB2 Design Review Guidelines
DB2: Maximizing Performance of Online
  Production Systems
Entity-Relationship Approach to Logical Data
  Base Design
ORACLE SQL*PLUS: A User's and
  Programmer's Tutorial
ORACLE: Building High Performance
  Online Systems
SQL/COBOL Application Design and
  Programming
Practical Data Analysis

## SYSTEMS DEVELOPMENT

Effective Methods of EDP Quality Assurance
Handbook of Screen Format Design
The Complete Guide to Software Testing
A User's Guide for Defining Software
  Requirements
A Structured Approach to Systems Testing
Practical Applications of Expert Systems
Expert Systems Development: Building
  PC-Based Applications
Storyboard Prototyping: A New Approach to
  User Requirements Analysis
The Software Factory: Managing Software
  Development and Maintenance
Data Architecture: The Information Paradigm
Advanced Topics in Information Engineering

## MANAGEMENT

Strategic and Operational Planning for
  Information Services
The State of the Art in Decision Support
  Systems
The Management Handbook for Information
  Center and End-User Computing
Disaster Recovery: Contingency Planning
  and Program Analysis
Techniques of Program and System
  Maintenance

## MANAGEMENT (cont'd)

The Data Processing Training Manager's
  Trail Guide
Winning the Change Game
Information Systems Planning for
  Competitive Advantage
Critical Issues in Information Processing
  Management and Technology
Developing the World Class Information
  Systems Organization
Project Management: A Comparative
  Analysis of PC-Based Systems
The Technical Instructor's Handbook: From
  Techie to Teacher
Collision: Theory vs. Reality in Expert
  Systems

## TECHNOLOGY

VSAM Techniques: Systems Concepts and
  Programming Procedures
How to Use CICS to Create On-Line
  Applications: Methods and Solutions
CICS/VS Command Level Reference Guide
  for COBOL Programmers
Data Communications: Concepts and
  Systems
Designing and Implementing Ethernet
  Networks
C Language for Programmers
Network Concepts and Architectures
SQL Spoken Here for DB2: A Tutorial
SQL for dBASE IV
Systems Programmer's Problem Solver
CASE: The Potential and the Pitfalls
Open Systems: The Basic Guide to OSI and
  its Implementation
An Introduction to Data and Activity Analysis
DOS/VSE/SP Guide for Systems
  Programming: Concepts, Programs,
  Macros, Subroutines

## THE QED INDEPENDENT STUDY SERIES

SQL as a Second Language
DB2: Building Online Production Systems
  for Maximum Performance (Video)

**For Additional Information or a Free Catalog contact**

QED INFORMATION SCIENCES, INC. • P. O. Box 82-181 • Wellesley, MA 02181
Telephone: 800-343-4848 or 617-237-5656

# THE TECHNICAL INSTRUCTOR'S HANDBOOK
## *From Techie to Teacher*

## Sue Redkey

QED Information Sciences, Inc.
Wellesley, Massachusetts

Library of Congress Catalog Number: 88-35467
International Standard Book Number: 0-89435-276-8

Printed in the United States of America
89 90 91 10 9 8 7 6 5 4 3 2 1

**Library of Congress Cataloging-in-Publication Data**

Redkey, Sue.
    The technical instructor's handbook : from techie to teacher / Sue Redkey.
        p.  cm.
    Bibliography: p.
    Includes index.
    ISBN 0-89435-276-8
    1. Technology—Study and teaching.  I. Title.
T65.R38   1989
607'.1—dc19

*Illustrations by Libby Barrett.*

*To my daughter Sara*
*and to my friend Art*

# Contents

*Foreword* . . . . . . . . . . . . . . . . . . . . xi

*Preface* . . . . . . . . . . . . . . . . . . . xiii

*Introduction* . . . . . . . . . . . . . . . . . .xv
   Being Technical and Being an Instructor . . . .xv
   *Using This Handbook* . . . . . . . . . . . xvi

*PART 1:  THINGS EVERY INSTRUCTOR SHOULD KNOW*   1

*About Learners* . . . . . . . . . . . . . . . . . 3
   What Are Adult Learners Like? . . . . . . . . 3
   How Do They Learn? . . . . . . . . . . . . 7
   Checklist for Teaching Adults . . . . . . . . 8
   How Well Do They Remember? . . . . . . . 8
*About the Learning Environment* . . . . . . . 11

*PART 2:  PREPARING TO TEACH* . . . . . . . . . . 17

*Know Your Subject* . . . . . . . . . . . . . . 19

*Know Your Course* . . . . . . . . . . . . . . 21

*Know Your Audience* . . . . . . . . . . . . . 25
   What Do You Need to Know? . . . . . . . 25
   How Do You Find What You Need to Know? . 26

*Getting Ready* . . . . . . . . . . . . . . . 29
  Preparation Checklist . . . . . . . . . . . 29
  Visualizing Success . . . . . . . . . . . 31
  Walkthroughs . . . . . . . . . . . . . 33

*PART 3: TEACHING TECHNICAL COURSES*    35

*Communication Skills* . . . . . . . . . . . . 37
  Components of Effective Communication . . 37
  What You Say—The Verbal Message . . . . 39
  How You Say It—The Nonverbal Message . . 42
  Changing Your Speaking Patterns. . . . . . 49
  How You Listen . . . . . . . . . . . . . 49

*The Presentation* . . . . . . . . . . . . . . 53
  Purpose of the Presentation . . . . . . . . 53
  Presentation Methods . . . . . . . . . . . 53
  Feedback/Evaluation Approaches . . . . . . 58

*Teaching Aids* . . . . . . . . . . . . . . . 61
  Why You Want to Use Teaching Aids . . . . 61
  Types of Aids . . . . . . . . . . . . . . 62
  Using Aids Effectively. . . . . . . . . . . 62

*Presentation Tips and Techniques* . . . . . . . 67
  Dealing with Nervousness . . . . . . . . . 67
  Getting the Class off to a Good Start . . . . 71
  Controlling Class Flow . . . . . . . . . . 74
  Making a Transition to the Next Topic . . . 76
  Effectively Teaching Technical Material . . . 78
  Team Teaching . . . . . . . . . . . . . 81
  Dealing with Problem Participants. . . . . . 83
  The Art of Asking Questions . . . . . . . . 88
  Responding to Answers . . . . . . . . . . 91
  Responding to Questions . . . . . . . . . 94
  When Your Answer is "I Don't Know" . . . . 98
  Conducting a Hands-On Workshop . . . . . 99
  Conducting Demonstrations . . . . . . . . 101
  Getting Physical . . . . . . . . . . . . 104
  Taking Care of Yourself. . . . . . . . . . 106
  Expanding Your Knowledge Base . . . . . . 108

Your Professional Presence . . . . . . . . 109
Ending the Class . . . . . . . . . . . . . 109

*APPENDICES*                                     113
*Appendix A:* Walkthrough Evaluation . . . . . . . . . 115
*Appendix B:* Tips for Vendor Instructors. . . . . . . . 119
*Appendix C:* Tips for Volunteer Instructors. . . . . . . 123
*Appendix D:* Networking. . . . . . . . . . . . . 125
*Appendix E:* Instructor's Diary . . . . . . . . . . . 139
*Appendix F:* Quick Reference Checklists . . . . . . . . 139

*Suggested Reading* . . . . . . . . . . . . . . . . 145

*Index* . . . . . . . . . . . . . . . . . . . . . 147

# *Foreword*

**P**erformance-minded managers are becoming increasingly aware of the importance of training and education as they re-shape and fine-tune their companies and organizations for the increasingly competitive decade ahead. Even the prestigious Harvard Business Review acknowledged this trend when, in a recent article (*Six Lessons for the Corporate Classroom*), they noted that *"Training is improvement for today; development is an investment for tomorrow."*

This new appreciation of the value of training also means the appearance in the corporate classroom of increasing numbers of line managers and staff; people who have no organizational connection to the Training or Human Resources departments but who augment and supplement training professionals in the classroom. The Harvard Business Review again: *"Company veterans make good teachers; they've been in the trenches."*

True, some companies have regularly included a tour of duty in the corporate classroom for "Hi-Pots," those fast-track professionals who are publicly acknowledged to be "going places." In these companies, experience in teaching is viewed to be a high priority developmental experience that "is an investment for tomorrow."

It's also true, however, that too few companies and organizations take advantage of the talent in their own ranks to leverage the skills of their professional trainers. "But they're only analysts (or designers, or . . . ) . . . they're not trainers" goes the familiar refrain. "And besides, they don't know anything about how the training process works. And we don't have time to teach them."

This book is the answer. *The Technical Instructor's Handbook* is a valuable reference guide for technical systems professionals who find themselves in the new role of instructor or facilitator. Sue Redkey writes about what she knows; she is continually working closely with non-training personnel who will be facilitating courses and workshops in addition to performing their regular responsibilities.

Training and educating adults is a privilege and I believe it is one of the most rewarding assignments a person can undertake. It is also a challenging and, yes, intimidating task, especially for a person who does not have a background in adult education or professional development. (I know—I've been in the "corporate classroom" for over 20 years and the more I do it, the more I respect those who want to learn to do it better.)

This is what Sue's book will do for you. It's really a "survival manual" for not only those preparing to teach but also for those who want to brush up on their techniques. She touches all the bases, from what you need to know about adult learners to how you conclude a course successfully.

Every instructor, full time or part time, should own a copy of this book. And use it. Often. Those who do will have taken a giant step toward mastering a challenging and most rewarding skill.

*Stewart L. Stokes, Jr.*
*Vice President*
*QED Information Sciences, Inc.*

# *Preface*

*T*he *Technical Instructor's Handbook* is written for technical professionals who also happen to be, or want to be, instructors. It is not a textbook. It is designed in an easy reference format that is intended to be used on the job or as a supplement to formal training in giving effective technical presentations.

While the primary audience is new instructors, experienced instructors, too, will find the easy-reference checklists in Appendix F useful as they prepare for and teach their classes.

# *Introduction*

## *BEING TECHNICAL AND BEING AN INSTRUCTOR*

The best technical instructors are not only strong in their technical areas of expertise, they are also skilled in giving effective presentations. It is rare to find this combination of talents occurring naturally in one individual, so developing outstanding instructors usually means showing good technicians how to teach, or taking good teachers and teaching them technical skills. This handbook is designed to aid you in becoming an effective technical instructor if you already possess the technical skills you will be teaching.

Here are some things you will want to remember as you make the transition from technician to technical instructor:

**1.** Knowing and teaching what you know are two different things.

**2.** Being a teacher is a different profession from being a technician; being both is yet another.

**3.** Strong teaching skills will usually make up for technical weakness more than strong technical skills will make up for a lack of teaching ability.

**4.** Teaching job-related classes to adults is very different from teaching children or college students.

**5.** Teaching, like learning, can—and should be—fun!

## *USING THIS HANDBOOK*

As stated above, technical (computer-related) teaching is a profession that represents a combination of two other professions: data processing or personal computing, and adult education.

This handbook is based on the premise that, as a technical instructor, you have knowledge and experience in the related technical profession but not in the education profession. It is intended to be used not as a self-study instructor training course but as a reference guide that can supplement formal training.

This section gives you a brief overview of the way the handbook is organized.

### *PART 1: Things Every Instructor Should Know*

Part 1 contains some basic information about adults as learners and about how different learning objectives dictate different teaching styles. These sections are not intended to duplicate the many books about adult learning theory, but rather to introduce the novice to these very important concepts and to provide some guidelines for teaching with the adult learner in mind. To be a successful instructor of adults, you must understand the adult learner.

Refer to Part 1 if you are new to teaching adults, especially if you have previously taught nonadult audiences.

### *PART 2: Preparing to Teach*

Part 2 addresses preparation for teaching. It covers background preparation (knowing your subject, your course, your audience), as well as more immediate preparation—getting yourself ready to teach and preparing the classroom.

Refer to Part 2 if you feel you need help in preparing to teach.

### *PART 3: Teaching Technical Courses*

The topics in Part 3 cover the things you want to know as you actually stand in front of a class and teach.

The first three sections provide some good background information, while "Presentation Tips and Techniques" offers suggestions on handling a variety of in-class situations, from getting over being nervous and handling problem participants, to making yourself physically comfortable while you teach.

Refer to Part 3 as a guide during preparation or as a handy reference during class. The sections are designed so that you can turn to whichever one you need without having to read the others.

### Appendices

As their titles imply, the appendices provide a sample walk-through evaluation form, some tips for vendor and volunteer instructors, some notes about networking, and quick-reference checklists.

Appendix E, "Instructor's Diary," follows a typical instructor through the preparation and presentation of a course she has never taught before. In these pages, which are based on fact, you will see how one instructor deals with a variety of typical issues.

### Suggested Reading

This is a list of books and articles that you may find interesting or relevant to your teaching career.

# PART 1

# *Things Every Instructor Should Know*

# About Learners

## WHAT ARE ADULT LEARNERS LIKE?

If you have never taught before, your image of what a teacher does may be based on the teachers you had during your school years:

> Teachers taught, students listened.
> Students were there because they were given no choice.
> Students had no say in what they were taught, or how or when.
> Students were expected to learn by memorizing.

Teaching adults doesn't go like that. Adults are different from children, and the ways we typically teach children, right or wrong, are not effective in teaching adults in most learning situations.

Here are some characteristics of adult learners, along with some guidelines based on these characteristics to help you be an effective teacher of adults.

### Self-Directing

Adults are generally in class because *they want to be there*, not because society requires it. This means that adult learners decide what they want to learn and when. They choose to come to class, and they can choose to leave class if it is not meeting their needs.

Of course, most of the adults you teach will probably be in class to learn concepts and skills required in their jobs. But even if they tell you they are there because "my boss told me to come," they still have choices. And it is generally not up to you to force them to stay in class. They, or their bosses, may decide at any time that they don't need to be there.

Self-directed adults respond well in a friendly, informal learning environment where they can take an active role and be given feedback about how they are doing.

## Teaching Implications

*Be aware of the learners' objectives* for being in your class: you are there to meet their needs, not the other way around.

*Inform learners of your expectations*, so there are no surprises. Be clear about class hours, breaks, homework, workshops, etc.

*Provide frequent reinforcement* for contributions and effort.

*Provide feedback* so learners can measure their progress against expected performance.

*Be supportive*, not judgmental; each learner will progress at her own pace, and your job is to cause learning to occur, not to rank learners against each other or against your idea of how quickly they should be learning. (Some learning experiences, such as entry-level training programs, may require you to judge or rank a learner's abilities, but in most cases, you will probably be expected to objectively evaluate, not judge.)

### Experienced

All adults have experience. It may not be in the subject you are teaching, but it is experience nonetheless. Adults' brains are not like blank slates waiting to be written on—they are full of life experiences, to which they will be adding what they learn in your class.

In addition, adult learners are generally eager to share their experiences with the class. That is one way they relate what they are learning to what they already know, and it is a great way for participants to learn from each other.

Be aware, too, that part of what makes a room of adults so interesting is that they all have *different* experiences; they are all unique. A group of 30-year-olds will have a lot less in common than a group of 10-year-olds. And one of their differences will be the ways in which they learn. (See "How Do They Learn.")

## Teaching Implications

*Tune in to your learners' experiences.* During opening introductions is a good time to find out about the relevant experience each learner is bringing to the class.

> *"Have you ever worked with a PC before?"*
>
> *"How many of you have used a database management system before?"*
>
> *"As you introduce yourselves, tell me what experience, if any, you have with on-line system development."*

*Show respect* for each learner's experiences. If a learner feels you don't value who he is (i.e., his experiences), he is not likely to be comfortable interacting in your class.

*Relate what you teach to those experiences.* Because adults do have so much knowledge and experience, they need to know where and how to fit in what you are teaching with what they already know. Analogies are helpful here:

> *"The stages of system development are much like the stages you go through in designing and building a house."*

Then develop the analogy, relating to what your audience probably knows about building a house.

*Build in a logical sequence.* Adults will be looking for logical relationships.

*Relate facts to each other*, so that learners can see the big picture.

*Pause after important points* for your learners to assimilate what you are saying with what they already know. This will also give them time to determine whether they have any questions.

*Encourage learners to share experiences.* This is a great way to increase audience involvement, and it allows learners to learn from each other and you to learn from them. This shared learning is one of the outstanding advantages of adult classes.

You can encourage sharing by beginning a guided discussion (see "Presentation Methods" or by allowing learners to share their comments as they indicate a desire to do so.

*Teach to different learning styles.* Because your audience will likely represent a variety of learning styles, you want to vary your presentation method periodically. This way each learner will have at least part of the class taught in the way she learns best. (See "How Do They Learn.")

### Ready to Learn

For adults to be motivated to be in class, they must be ready to learn. This usually means they want to learn, or have a need to learn, this subject at this time. Despite this want or need, many adults will be quite set in their ways, and this can present a challenge to you.

While each learner will have a preferred way of learning, most adults will respond to a variety of teaching methods. This means that there are a variety of ways to teach them effectively, and that, as learners, they like variety.

### Teaching Implications

*Do not force attendance or involvement.* People will take part when they are ready to.

*Show respect for the old ways* that your learners are familiar with.

*Use a variety of teaching methods.* This will make the class more interesting for everbody and will help to hold your audience's interest.

### Problem-Centered

Adults are often in class to solve problems: to learn something that will make their jobs easier; to qualify for a promotion; to prepare for an upcoming project. If they have any experience already with the subject of your class, they will likely have questions based on real situations they have encountered. And they will be expecting your class to give them the solutions they need.

Because they are problem-oriented, they will have little interest in topics they see no direct application for. Generally, they will be applying tomorrow what they learn today.

### Teaching Implications

*Provide realistic examples and exercises.* This will give learners an opportunity to see how what you're teaching can be applied to their real situations.

*Provide hands-on practice* wherever possible, so that learners come away from class not just "knowing how," but "having actually done."

*Encourage questions* and opportunities for learners to share problems and experiences.

### ༼ᵧ General Tip

You yourself are an adult learner: when in doubt, think about what you like and don't like when you are in a technical class.

## HOW DO THEY LEARN?

Being an effective instructor requires some understanding of how people learn. This book will not try to duplicate the many books that explore this subject, but it will explain the major differences that you want to be aware of. As an instructor, you want your course designs and presentations to reflect your awareness of these differences.

This doesn't mean that you want to teach each course a dozen different ways, but it is one of the reasons you want to vary your presentation method, giving each class a mix of learning experiences. This way, hopefully, everybody gets at least some of what he really likes best.

Here are some observations that will help you understand your adult audience.

**1.** Some people learn new information best if it is presented sequentially; others prefer to get the big picture first and then proceed in a top-down fashion.

**2.** Some learners prefer to work in groups, while others perform better if they can work independently.

**3.** Some people learn faster those things that they see or read; some learn faster what they hear and discuss; and others learn best what they experience.

**4.** Some learners want to be told (or shown) everything you want them to know; others like to be led and then be allowed to figure out the rest for themselves.

**5.** Some people need to take detailed written notes while they are learning, while others are able to take notes mentally.

**6.** Some learners take learning very seriously and expect the whole class to be strictly business; others approach learning as fun and expect to—and generally do—have a good time in class.

**7.** Some adult learners need special encouragement in order to help them relax and be receptive to learning, to build their confidence, or to get them to be comfortable contributing in class.

**8.** Most adults learn best in a comfortable, casual learning environment.

## ⌇ *CHECKLIST FOR TEACHING ADULTS*

The following checklist is a quick-reference guide to the many things to keep in mind as you teach adults. Each of these points is discussed fully elsewhere in this handbook.

**1.** Be flexible in your teaching style and in response to your learners' needs.

**2.** Let your audience know what to expect.

**3.** Tune into your audience's needs.

**4.** Show respect for your participants and their experiences.

**5.** Vary your presentation methods.

**6.** Use visual aids.

**7.** Get your audience involved.

**8.** Provide hands-on practice.

**9.** Allow for questions and discussions.

**10.** Allow time for participants to take notes.

**11.** Allow for breaks every hour or so.

**12.** Use realistic examples.

**13.** Build information in a logical sequence.

**14.** Provide meaningful transitions between topics.

**15.** Make all assignments meaningful.

**16.** Have fun!

## *HOW WELL DO THEY REMEMBER?*

As we have seen, there are different ways that people learn and, therefore, different ways of teaching effectively. It helps to be aware, too, that regardless how you teach, much of what your audience learns is eventually forgotten.

Naturally, the more we use, the more we are likely to remember. But the way the material is presented also has an effect on our ability to retain information—even just 48 hours later— as the following table illustrates.

| If we Learn by . . . | We may retain . . . |
|---|---|
| hearing | 10–15% |
| seeing | 15–30% |
| saying | 30–50% |
| doing | 50–75% |
| doing with coaching | 75% |

## Teaching Implications

It is safe to assume that much of your teaching will involve verbal delivery of information; that is, your audience will be learning by hearing.

Since you know learners are likely to forget much of what they only hear, you will want to enhance their learning experience by providing them the opportunity to see, say, and do what they are learning. This means including things like the following.

*Use visual aids.* Not only is seeing plus hearing more effective than hearing alone, but visual aids help explain complex ideas (a picture is worth a thousand words), and they make the presentation more interesting, increasing the audience's attention span. (See "Types of Aids.")

*Have oral reviews,* giving learners the opportunity to repeat out loud what they know. The process of integrating information mentally before expressing it verbally can give learners confidence in their ability to remember and pull together related pieces. The act of saying it out loud helps them retain what they have learned. (See "Feedback/Evaluation Approaches.")

*Conduct guided discussions.* This is another way to encourage learners to say out loud what they know. (See "Presentation Methods.")

*Include exercises and workshops.* Give learners the opportunity to do what they are learning. The form this practice takes will vary depending on the subject, but generally, we are talking about written exercises and hands-on workshops. (See "Feedback/Evaluation Approaches" and "Conducting a Hands-On Workshop.") *Be available to coach* during these exercises and workshops, where possible, to provide the maximum learning benefit.

In general, the more often you can let learners do, review, and talk about what they have learned, the more they are likely to retain. And each time, that is good feedback to you about how well they understand what they are learning.

# *About the Learning Environment*

*A*dult learners generally respond well to a learning environment that is pleasant, comfortable, and informal. While you as the instructor may not have direct control over all aspects of the classroom environment, you definitely control or influence many of them.

The total learning environment consists of both the physical environment and the psychological atmosphere. Here are some tips that can help you provide an environment in which your participants enjoy a meaningful learning experience.

### *Physical Environment*

Try to use a classroom that provides basic physical comfort to the participants and to yourself. Don't be shy about asking for adjustments where feasible. You want to pay attention to things like

> Comfortable chairs
> Adequate table space
> Adequate leg space
> Appropriate lighting
> Comfortable temperature
> Outside noise

Another aspect of the physical environment is the overall room arrangement. How your classroom is arranged can affect not only the comfort and expectations of your audience, but also the success of your class. You want the room arrangement to:

Allow all participants to see you and any visual aids you are using.
Encourage interaction among participants.
Facilitate any group activities you have planned.
Allow for easy exit in case of emergency.

The advantages and disadvantages of several typical styles of room arrangement are discussed below and illustrated in the accompanying diagram. If you have no say in the arrangement of the room, at least you can be aware of the possible limitations of the arrangement you are working with.

*Theater style*: In this arrangement, there are no tables and the chairs are arranged in rows. This kind of arrangement does not make it easy for people to take notes, work with handouts and manuals, or interact with each other. It is best suited to a brief lecture presentation rather than a technical class. Seats should be arranged with sufficient aisles to make prompt exit possible in case of emergency.

*Traditional classroom style*: This arrangement consists of tables and chairs, usually with two or three people at a table. All the tables face directly forward. If there are terminals or PCs in the room, they may be on the tables in front of the participants, or they may be in a separate area off to the side or back of the room. Classroom style does give participants room to spread out their materials, but it is not the best arrangement to encourage interaction among the participants. As with theater style, this arrangement should include sufficient aisles to facilitate exit.

*Horseshoe*: A horseshoe consists of tables and chairs arranged in a horseshoe or "U" shape. This arrangement has the advantage of allowing participants to see each other face to face, which encourages interaction. There is also table space for participants' materials. You seldom see this arrangement with terminals or PCs right in front of the participants because of the wiring required. Rooms with this kind of equipment usually have the equipment in a permanent location, and a horseshoe is seldom a fixed arrangement.

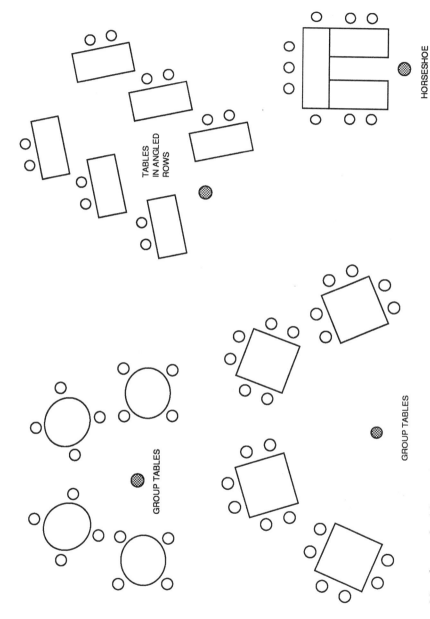

TABLES
IN ANGLED
ROWS

HORSESHOE

GROUP TABLES

GROUP TABLES

Numbers of tables and seats can vary, but, regardless of class size, everyone should be able to easily see the instructor and other participants, if successful interaction is to be achieved.

*Angled rows*: This arrangement is a variation on traditional class-room style. If the room is arranged with tables on both sides of a center aisle and there is room to turn them at an angle, the angle will allow the participants to see each other better and encourage interaction more than with classroom style.

*Group tables*: In this arrangement, each participant sits at a (round or square) group table. Ideally, everyone can face front easily when focused on the instructor and can easily interact with others in the group around the table. This kind of arrangement is best for those classes in which working in groups is a major part of the learning experience. Round tables facilitate interaction best, and square or rectangular tables are easiest to arrange so that every participant can face front easily and still use the table surface for taking notes.

*Break-out Rooms*: These are extra rooms that participants may use during group workshops. If your room cannot be arranged with group tables, or if each group must have quiet space in which to interact, using break-out rooms is recommended. Ideally, they are directly adjacent to the main classroom.

Remember that if your participants are not physically comfortable, they will have a hard time concentrating on the class; and if they do not have adequate work space, they will, at best, have a frustrating experience.

### Psychological Atmosphere

If people are not psychologically comfortable in your class, their learning experience can be severely limited. The psychological atmosphere of your class is mostly up to you—regardless what attitudes, fears, or political concerns participants bring to class, you are the one who sets the tone once everyone is together.

It is important that you let participants know they are free to:

Ask questions
Make comments
Share experiences
Make mistakes

The first thing you can do is to announce at the beginning of the class that you encourage participation. The most important thing you can do is to *demonstrate* that people can feel comfortable contributing and being wrong.

Always show respect for your participants and the comments and ideas they contribute.

Build on what they contribute by incorporating their ideas and experiences into your lecture and examples.

Respond to all questions and comments.

Encourage mistakes in workshops, reminding learners that the more mistakes they make, the more they learn.

Probably the best advice here is to put yourself in the place of your audience and think how *you* would like to be treated.

For more ideas on responding to questions and encouraging your audience to interact, refer to "The Art of Asking Questions", "Responding to Answers" and "Responding to Questions."

# PART 2

# *Preparing to Teach*

# Know Your Subject

**O**bviously, when you are the instructor, people expect that you know what you are talking about. You probably have similar expectations.

But there may be times when you have to teach and you *don't* really know the subject all that well; that can be one of the hardest things to do as an instructor. If this has ever happened to you, you know the risks you run:

> You have a hard time answering questions.
> You have trouble fully explaining some points.
> You leave things out.
> Fairly basic problems come up in workshops, and you've never seen them before.
> You lose your confidence and, possibly, your credibility.

One way to avoid these risks is to know your subject as thoroughly as you can *before* you teach. This doesn't mean knowing everything there is to know—if you wait for that kind of knowledge, you will never make it to the classroom. Besides, in this field, you know that constant change makes it impossible to know everything anyway.

Short of that unrealistic goal, here are some things you can do to achieve a high level of confidence with your subject.

**1.** *Learn more than you expect to teach.* When you are teaching others, it helps to have some depth; you never know what kinds of questions will come up.

**2.** *Get hands-on experience*—all you can. Ideally, if you are teaching database design, you will have designed a real database first; and if you are teaching PC DOS, you will have used all the commands you are teaching.

**3.** *Learn from others what you cannot experience.* If you are teaching LOTUS to a group of managers who want to use it to keep track of budgets, and you have never done a budget application yourself, try to find an example of one. Find out the important points your audience is likely to bring up.

**4.** *Stay current.* This is a biggy. If you are a full-time instructor, you know how hard it can be to stay current in the subjects you teach. There is no way you can both "do it" full-time, and "teach it" full-time.

If you are an occasional instructor, you are much more likely to stay current technically, but your teaching skills may lose their edge.

Either way, this one is a challenge. The key is that no matter how much time you spend teaching, it is best to spend *some* time "doing."

*Use the skills you teach.*
*Network with others* who use those skills in their jobs.
*Keep the software you use updated.*
*Keep updated manuals.*
*Take refresher courses.*

**5.** *Know who the experts are*, whether they are in your company or in other companies. If you anticipate needing help during class, find out who can be available to answer questions should you have any. It is probably a good idea to add to your instructor notes a list of experts and their phone numbers. Remember to let your contact(s) know that you are teaching and may need to call.

# Know Your Course

*I*n addition to knowing your subject, you also want to thoroughly know your course. Assuming you can teach spreadsheets just because you have worked with them for years is like assuming you can play Hamlet just because you know Danish history.

Here are some typical problems that can arise, and the kind of preparation that can prevent them.

*You have no idea how long it will take to teach the class.*

Have a walkthrough, practice your lectures, *talk to others* who have taught the course before. Then, as you teach the real thing, make notes about the time you spend on each section of the course, how long workshops take, etc. Do this for the first several times you teach the course, and you will get an idea of average times. It also helps if you allow yourself to become comfortable with the fact that your timing will not be perfect the first time you teach a class.

*You get sidetracked with questions* that get into areas you hadn't planned on teaching.

Know the scope of the course. Know what it does not cover, as well as what it does cover. This way, you will know before you begin to answer a beyond-scope question that it is one you want to answer outside class time.

*You get into side discussions of topics that come up later*. You may get going in answering a question, not realizing that you will be covering that topic a half-hour from now anyway. You will appear poorly prepared if you don't know what's coming.

Know the flow of the presentation. Again, a *walkthrough* is an excellent way of preparing and can help you learn the order in which things flow. Also, *instructor notes* can help you stay on track; you may even have a note that says "DO NOT talk about (whatever) here," if it is something that typically comes up prematurely.

*You go into way too much detail or too little detail.* Usually, if you know a subject really well, the tendency is to go into too much detail. Besides throwing off your timing, too much detail can overwhelm a class that was not expecting that depth.

Know the depth of the course. Know before you go into the classroom how much depth the course calls for. Sometimes the appropriate depth will vary depending on the audience, which is one more good reason to know your audience.

*You can't remember the meaning of a visual aid.* This is especially likely to occur when you are not the one who developed the course.

Know each visual aid thoroughly before you use it. No fair saying "I don't know what (whoever designed it) had in mind here, but . . ." Besides being unprofessional, that kind of remark can hurt your credibility. If you are stuck with something you really cannot explain well, don't use it. Redesign it or do without it.

*You can't explain some of the things that come up in workshop.* You are frequently saying things like "I've never seen that before," or "That shouldn't have happened; try it this way, instead of the way the instructions said," or "Oh, yeh, I meant to tell you—when you hit the attention key right there, you lose everything."

Practice every exercise and workshop before class. Even if you have taught the class before, you want the details to be fresh in your mind. You may be working with a new release of the software, so you need to know what to expect. You want to remember the little things to tell the class about, and you want to remember the typical problems learners encounter and how to resolve them. Exercises and workshops should not appear as new to you as they do to your audience.

Keep current with new software releases so you know how they will affect what you teach and what to expect in hands-on workshops.

*You have questions about the design or flow of the class.* This could include anything mentioned above; it could include questions that come up during your preparation or questions that come up while you are teaching.

Know who has taught the course before and, if possible, who developed it. Keep names and phone numbers with your instructor notes, and call on these people if you need to for clarification.

Over time, you will get to know your courses very well, but by following the above suggestions, you should be able to avoid some of the most damaging pitfalls even the first time out.

Knowing your course thoroughly is like knowing the script to a play, although, unlike a play, a live course will be different every time. There will always be the things you can't predict—all the more reason to be prepared for the ones you can.

# *Know Your Audience*

## *WHAT DO YOU NEED TO KNOW?*

One of the most important parts of your preparation is learning about your audience. You want to understand the general audience that the course is designed to address; and you want to prepare for—and possibly tailor to—the specific audience you will be teaching this particular time.

Understanding the general audience will help you identify the appropriate scope and depth for the class. Knowing your specific audience will help you set the pace for your presentation, select appropriate examples, relate to your audience's previous experience, and possibly tailor parts of the class to meet this audience's needs.

Here are the kinds of things you want to know—before the class if possible—about both your general and your specific audience.

> *Who are they?* Are they technical people? Are they end users? Are they managers or executives? Are they fellow employees? Are they clients?
>
> *Where are they from?* What division(s) do they work in? Are they from your company or from other companies?
>
> *Why are they attending your class?* Are they looking for an overview? Are they expecting to learn skills? Do they just want a refresher?

*How will they use what they learn?* Will they be using it right away? Is there a specific project to which they will apply this knowledge? Will they be using the skills themselves? Or managing others who use these skills?

*What experience, if any, do they already have?* Who in your audience is most experienced in the topic you are teaching (or related topics)? Who is least experienced? What related experience do they have?

*In what environment will they use what they learn?* Are there constraints (technological, political, etc.) that will affect how you teach the class, or how knowledge/skills from the class will be applied?

## HOW DO YOU FIND OUT WHAT YOU NEED TO KNOW?

How do you learn all these things you want to know about your audience?

### General Audience

Assuming the course is already developed, the audience for which it is intended is probably documented with the course materials. If it is not, you can probably get an idea by looking at the course materials. For instance, if you can't tell from the course title ("VSAM" or "LOTUS 1-2-3") whether the course is geared toward beginners or experienced users, a look at the content should eliminate any confusion.

If you are developing the course yourself, you will usually define the intended audience based on input from those who requested the course.

### Specific Audience

To learn about the specific audience you are teaching now, here are some suggestions.

Have participants introduce themselves at the beginning of the class. Be sure to ask for the information you need; don't leave it to chance that they will offer it. It is a good idea to list on the board or on a flip chart what you want people to tell you. For example:

Your name
Your job title/description
What experience you already have with the course subject
What you hope to get out of the class

You may want to contact participants prior to class if there are things you need to know in order to prepare to teach. For example, if you have participants from several different divisions or offices where things may be done very differently, you might need to know ahead of time exactly what they are expecting. Anytime the course needs to be tailored, you need to know your audience ahead of time.

If you are giving a presentation to a homogeneous group, you may be able to find out all you need to know by calling the group's manager to learn about the group and their expectations (and their manager's expectations). This approach may be helpful not only from a teaching standpoint but also from a political one, especially if the manager has requested the class.

No matter what other methods are used, you can always benefit from having participants introduce themselves at the start of the class. It will help people get to know one another, and the interaction helps everyone relax.

There will always be unexpected things that can come up when you are teaching, but by knowing your audience, you can eliminate some of the worst surprises.

# Getting Ready

## PREPARATION CHECKLIST

The following checklist suggests things you can do to get ready to teach a class.

### Preparing Yourself

1. Know the subject you are going to teach.
2. Know the course you are going to teach.
3. Know your audience.
4. Visualize success.
5. Be well-rested.
6. Be set up and ready before participants arrive.

### Preparing Your Presentation

1. Have a walkthrough.
2. Practice especially your introduction.
3. Tape record your lecture to hear your speaking style.
4. Work through all excercises and workshops.

## *Preparing Your Materials*

**1.** Have all your references ready.

Instructor guide
Sample solutions
Technical manuals

**2.** Have all your teaching aids ready.

Transparencies—clean and in order
Blank transparencies if you use them
Flip charts—prepared or blank
Any other visual aids
Markers—make sure they all write
Pointer

**3.** Have all your participant aids ready.

Handouts—in order and enough copies
Name cards if you use them
Roster
Evaluation sheets

## *Preparing the Classroom*

**1.** Arrive early—about an hour on the first day, maybe 15 to 20 minutes on subsequent days.

**2.** Arrange the room the way you want it.

**3.** Make sure all equipment you need is there.

**4.** Make sure all equipment is ready to use.

Projector is correct distance from screen.
Projector works and is focused.
Screen monitor works and is focused.
Flip chart has enough paper.
You have the markers you need and they all write.
System is up/PCs are booted.

**5.** Know where light switches are and how they work if there is anything unusual.

**6.** Arrange the instructor's table the way you want it.

**7.** Be sure there is a clock or you have a watch with you.

**8.** Set out materials.

Handouts
Name cards
Evaluation forms (if you hand them out at the beginning of the
    class)

**9.** Be ready before participants arrive—usually 15 minutes before the class starts.

### Miscellaneous

**1.** Know class location.

Building
Room
How to get there
Who has the key

**2.** Know class hours.

**3.** Know who to call in the event of technical problems.

## VISUALIZING SUCCESS

One of the most positive things you can do for yourself in preparing to teach is to visualize yourself being successful.

One way to do this is to have a positive outlook whenever you encounter an obstacle or a problem and tell yourself that you know you will be able to handle it. Another way is to *not* give in to the temptation to imagine the worst.

Still another thing you can do is to practice creative visualization—taking an active role in visualizing your success. If you have never tried this technique before, the following steps will help you get started.

**1.** Find a place where you can be undisturbed for 20 to 30 minutes.

**2.** Sit comfortably with your back straight, either in a chair or on the floor. Be sure your clothes are not restricting your breathing.

**3.** If you like, play some relaxing music. The music will help you focus and keep you from being distracted by other noises.

**4.** For best results, close your eyes. This will help keep you from being distracted by visual stimuli.

**5.** Take several really deep breaths. Each time you breathe in, feel your body relaxing; each time you breathe out, release any tension you feel.

**6.** Allow your body to relax. If you like, take a few moments to visualize each part of your body relaxing, working either up from your feet or down from your head.

**7.** Now focus your mind on your teaching. Get an image in your mind of—visualize—yourself successfully preparing for and teaching your class.

**8.** See yourself, as if watching a movie, go through all the steps you need to go through to prepare for teaching: everything from writing instructor notes, photocopying materials, and holding a walkthrough, to getting enough sleep the night before and setting up the room.

**9.** Anytime you encounter a problem while you are watching—for instance, if you see yourself feeling nervous during the walkthrough—see yourself handle the situation the way you would really like to handle it. Allow yourself to feel relaxed as you overcome any problems or obstacles.

**10.** Repeat steps 8 and 9, visualizing yourself teaching the class.

**11.** Take as long as you like and feel free to repeat parts that you think need work.

**12.** When you are finished with your visualization, make sure you are still feeling relaxed, and gradually come back to your physical surroundings.

**13.** As you open your eyes, pause for a moment to notice how relaxed—and possibly energetic—you feel.

**14.** Visualize as often as you feel the need.


This kind of visualization is a way of programming yourself for success; it is like a dress rehearsal. When you go through the actual activities of preparation and teaching, it will be as if you are following a script, as if you've done it all before.

"Seeing" yourself deal with problems successfully can give

you the confidence to actually deal with them that way when they come up.

Give it a try, and good luck!

## *WALKTHROUGHS*

Walkthroughs are especially valuable if you have never taught before, or if you are teaching a course for the first time. Several of your preparation objectives can be accomplished with a walkthrough of your presentation. A walkthrough can help you

> Become familiar with course flow.
> Determine timing.
> Anticipate questions.
> Test exercises and workshops.
> Refine your presentation skills.
> Get feedback from others about the course and about your presentation.

If you've never done a walkthrough before, here are some guidelines to help you be successful. The assumption here is that this is a walkthrough of your presentation, not so much your course design. If you are designing the course, you will want to have a separate design walkthrough.

**1.** Try to use the room you will actually teach in for your walkthrough.

**2.** Schedule it about two weeks—or longer—before your actual class. This way you will have time to make necessary revisions that come out of the walkthrough.

**3.** Invite a few people (3 to 4 is a good number) to observe the walkthrough and give you feedback. Your audience should include people who can evaluate presentation skills and, if possible, at least one person who knows the content.

**4.** Prepare as you would for your class in terms of having visual aids and handouts ready.

**5.** Walkthrough only a portion of the class. If you really need to practice and get feedback on more of the class, hold several walkthroughs.

**6.** A good timeframe for a walkthrough is about 2 hours. This gives you time to present for an hour or so, take a break, and then hear the feedback from everyone in the audience.

**7.** Encourage your audience to use a written evaluation form to give you feedback. There is a sample walkthrough evaluation form in Appendix A.

**8.** Accept comments gracefully—both the praise and the constructive criticism. Remember that any problem you turn up in a walkthrough is one less problem to surprise you during the actual class.

**9.** Take notes, or have someone else take notes for you, as you receive verbal feedback.

**10.** After the walkthrough, make necessary revisions to visual aids, exercises, workshop instructions, and your presentation techniques.

# PART 3

# *Teaching Technical Courses*

# Communication Skills

## COMPONENTS OF EFFECTIVE COMMUNICATION

As an instructor, you are a professional communicator. However, it does not necessarily follow that you are a skilled communicator.

Some knowledge of what constitutes effective communication along with practice of your communication skills can help you be a more effective instructor.

First, let's take a look at the components of effective communication.

### Sending

The first step in a communication cycle is the sending of a message. Since our context here is classroom instruction, we will assume that we are talking about sending a voice message, or speaking.

The speaking message that is sent consists of two major components:

The verbal message—the words you say (see "What You Say")
The nonverbal message—*how* you say the words (see "How You Say It")

## *Receiving*

For communication to occur, a sent message must be received —in this case, heard. Good listening skills are helpful here for the receiver. In addition, the receiver will be interpreting *how* you say what you say, and that interpretation—more than your meaning—will determine how successfully you have communicated.

## *Getting Feedback*

Finally, you cannot be sure that you have *effectively* communicated until you, the sender, receive feedback from the receiver. Usually, this is in the form of a message the receiver sends you (spoken or nonverbal). Then you become the receiver, which implies that you will need good listening skills yourself.

Here's what all this means to you as an instructor:

Talking for six hours straight is not necessarily communicating.

You will need to get feedback from your audience to know whether or not your communication efforts are successful (see "Feedback/ Evaluation Approaches").

In addition to developing both verbal and nonverbal communication skills, you will also want to develop listening skills.

## *What Are You Trying to Communicate?*

If you are the instructor, then obviously part of what you want to communicate is the information people have come to learn. But there are other things you want to communicate as well:

Respect for your audience.

Authority to be leading the class.

Control over the flow of the class.

Credibility in the subject you are teaching.

Confidence in yourself and in the course.

Psychological openness, so learners will feel free to ask questions and participate.

Keep this list in mind, and ask yourself these questions as you read through the next few sections:

Which of these things do you want to communicate?

How well does your speaking style communicate them?

Are there aspects of your style that you would like to change?

## *WHAT YOU SAY—THE VERBAL MESSAGE*

The words you speak are the verbal part of your communication. While you may think that your whole message is in your words, in fact, the words account for only about 7 percent of the total message you communicate.

Naturally, the message you send includes the words you choose to convey your meaning. When you are teaching, it is important that what you say is accurate, and it helps to be concise. That is, make sure you are telling people the right thing, and don't overwhelm or confuse them with extra words.

Something else you want to be aware of is that the use of certain words or phrases can work against you in communicating those things listed earlier (confidence, authority, etc.). Here is a list of the kinds of words and phrases that can rob you of a confident, in-control presentation.

*Hedge words* are words that make you sound tentative, as if you are not sure of yourself, or as if you are afraid of your audience. If they are used to convey their literal meanings, they do just that; but if they are used as extra words—filler—they work against you.

| | |
|---|---|
| I think | Just (as in "only") |
| Sort of | A little |
| Try | I guess |
| You know | Would you like |
| Maybe | Would you mind |
| I tend to think | Probably |
| Pretty (as in "very") | |

For example, instead of

> *"It's* probably *time for* a little *break.* Would you like *to take* just *a few minutes to* sort of *stretch and* maybe *get a cup of coffee? I* guess *we should* try *to be back in* about *10 minutes."*

Try

> *"I can see you're ready for a break. Let's take 10 minutes and be back at 10:05."*

The first example hedges all over the place, while the second example is direct. The second speaker clearly communicates more confidence and control. His class is much more likely to respect his authority and be back promptly from the break.

*Tag questions* are questions that we sometimes "tag" onto the end of our sentences. Again, if you really want to know what you are asking, that is one thing; but if the question is a rhetorical one that is tagged on out of nervousness or uncertainty, it can rob you of confidence and authority. If a tag question comes at the end of an otherwise authoritative statement, it has the effect of taking back what you just said.

|                   |                   |
|-------------------|-------------------|
| Don't you think?  | Right?            |
| Isn't it?         | OK?               |
| Don't you?        | Wouldn't you say? |
| You know?         |                   |

For example, instead of

> *"Whenever you use this parameter, you've got to be EXTREMELY careful, OK?"*

Try saying the same thing, only end with "careful!" instead of "OK?."

*Intensifiers, diminutives and rave words* are words that we often use when we're excited or when we're trying to please. Be aware that they are construed by many to be more appropriate in a personal conversation than in a business discussion, so you run the risk of sounding unprofessional if you use words like these too often. This does not mean you don't ever use them—they can add some welcome color to your speaking style.

| *Intensifiers* | *Diminutives* | *Rave Words* |
|---|---|---|
| So | Little | WOW |
| Very | Bitty | SUPER |
| Really | Bitsy | AWESOME |
| | Tiny | GREAT |
| | Teensy | |

For example, instead of

> *"I think you'll really, really like this great new little keyboard! It's really super!"*

Try

> *"I think you'll like this new keyboard. The keys have a light touch, and there are separate PF and function keys, so you need only one keystroke now for PF keys."*

On the other hand, this sounds fine:

> *"I just heard about your promotion—that's really great!"*

### Hear What You Say

Until you hear yourself on tape, you may have no idea how many of these words and phrases you actually use, and how often.

Try listening to yourself by taping a presentation or taping your end of a phone conversation. If you hear patterns you would like to eliminate, try to catch yourself, focusing on just one pattern at a time.

You might ask a friend to point out every time you say "you know," or "um," for example. Better yet, make a game out of it: go to lunch with a friend and each of you keep track of how many times the other says something she is trying to eliminate. Then, whoever catches the other the most gets a free lunch!

Over time and with practice, you'll find the unwanted words disappearing from your speech.

## *HOW YOU SAY IT—THE NONVERBAL MESSAGE*

93 percent of what you communicate through speaking comes from your nonverbal message. That means your verbal message —the words you speak—accounts for only 7 percent of what your audience receives. Effective communicators, then, are good not only at what they say but also at how they say it.

At times, the verbal and nonverbal messages do not match. As a professional communicator, you should be aware that when this is the case, those receiving the message are far more inclined to believe the nonverbal than they are the verbal.

Nonverbal communication can be either vocal—pertaining to the sounds you make as you speak—or nonvocal. These different categories are described in the following sections, along with some tips on effective nonverbal communication in a teaching situation.

### The Vocal Message

Your vocal message comes from how you *sound* when you speak. It makes up about 38 percent of the total message you communicate and includes these components:

> Pitch
> Volume
> Speed
> Voice quality
> Articulation
> Paralanguage

*Pitch* refers to how high or low your voice sounds. When we talk about how pitch affects our communication, we are generally looking at these three things:

> *Base pitch*—how high or low your typical speaking voice is. Audiences seem to prefer a pleasantly low base pitch; people with high-pitched voices tend to have a harder time commanding attention and communicating authority and control.
> *Intonation*—how much variation in pitch you use when speaking. A widely varying pitch can communicate emotion, excitement, and enthusiasm; a very even pitch becomes monotonous and

suggests boredom. As an instructor, you definitely want to avoid the monotone; you will be much easier to listen to if you have a medium range of variation as you speak, with an occasional wide range to generate enthusiasm or stress a point.

*Terminal pitch*—whether your pitch rises or falls at the ends of sentences. A rising terminal pitch suggests a question, but many people end nonquestioning statements with a rising pitch. This can be interpreted as your asking for permission or approval, and it robs you of your authority.

Notice how you say your name when you introduce yourself; most of us are inclined to make our introduction sound like a question. Try dropping the pitch at the end of this sentence: "Good morning. My name is _____." and see how different it sounds from this: "Good morning. My name is _____?"

A falling terminal pitch sounds authoritative and emphatic and should be used especially when you want to stress a point or communicate that you are in charge.

*Volume* refers to how loudly you speak. In front of a class, of course, you want to be heard, so you are probably conscious of trying to speak up as you teach. However, many of us have faulty internal volume controls and may think we are talking much louder or softer than we actually are.

The best way to know for sure is to get feedback from others and to practice. A tape recorder is helpful here, but you also need to have people listen to you and judge the volume.

People who speak with a very low volume may be perceived as weak or lacking in self-esteem, confidence, and credibility. Those characteristics are inconsistent with the image of authority and control you wish to project while teaching, so if you know you are a soft speaker, you may want to practice increasing your volume.

*Speed* is the speed of your delivery. People who speak very rapidly may be nervous, self-conscious, or maybe just unaware that speaking in front of a class is more effective at a moderate speed. It's true that the brain can process information faster than it can be spoken, so if all you need to do is understand the words that are being said, rapid speech presents no problem.

In a learning situation, however, people need time to assimilate what they are hearing, understand it, and determine if they have any questions. For instructing, a speed of about 150 to 160 words per minute is recommended.

Remember not to slow down too much, or you'll put people to sleep.

A good way to work on speed is to select a paragraph of 150 to 160 words, and time yourself as you read it into a tape recorder. Keep practicing until you fit the reading into one minute, and then notice how that sounds and what it feels like to talk at that speed.

*Voice quality* refers to the resonance of your voice: is it orally resonant or nasally resonant?

People generally find an orally resonated voice more pleasant to listen to than a nasally resonant voice. As an instructor, you certainly do not want your audience distracted and turned off because of a nasal twang.

The natural nasal sounds in English are the sounds of *N, M* and *NG*; however, we often make *A*s and *O*s nasal unnecessarily. The way to avoid the nasal sound is to open the back of your throat, so that the air passes through your throat rather than your nose as you speak.

Here is a list of words you can practice saying on tape to hear how they sound:

| | |
|---|---|
| Grand | Loud |
| Back | Cancel |
| Shout | Pat |
| Now | Clamp |
| Can't | How |
| Down | Clamor |

*Articulation* is the clarity with which you pronounce your words. Obviously, you want the words you say to be understood, and listening to yourself on tape will give you a good idea of whether they are or not. Your goal is to be distinct and clear without being so precise that you sound stilted.

There are certain words that we all tend to slur, and each of us has our own list of words and phrases that we slur, usually because we say them so often that we tend to speed up and get sloppy. For example, notice sometime how often in our field you hear people say "day prossing" instead of "data processing."

Here is a list of frequently used words that are often slurred, including some *L* words—words with an L in the middle or at the end that many people have trouble articulating clearly.

| | |
|---|---|
| All | A lot of |
| Critical | Supposed to |
| Did you | Children |
| Going to | Commercial |
| Have to | Let's |
| Filled | Did you know |
| Glasses | Want to |
| Couldn't you | |

*Paralanguage* is a category of sounds that are not real words, and have no literal meaning. Overuse can be distracting to listeners, and it can also make you sound unprepared or nervous. Even when that is the case, you don't want everyone to know, so these are sounds you could eliminate from your speaking style altogether and never be the worse for it.

Typical paralanguage includes

| | |
|---|---|
| Grunts | Ers |
| Sighs | Ahs |
| Ums | Uhs |

There is probably no need for an example here. Listen to yourself on tape and ask your friends to help you identify where you have a problem using these sounds. Then practice leaving them out. A silent pause is preferable to a nonword.

### The Visual Message

If people can see you as you speak, the visual message they perceive becomes the dominant portion of the message you communicate. How you look, how you move, and what you do with your eyes all contribute to the message.

Entire books have been dedicated to interpreting the visual message, and this section does not attempt to duplicate all that has been printed elsewhere. What it does do is highlight some of the things you want to be aware of as an instructor.

### Appearance

Your appearance says a lot about you: your clothing (style, color, quality, neatness, fit, appropriateness to the culture and the occasion); your hair style; your accessories; and what kind of makeup you wear, if any. All these things give people an impression of the kind of person you are; they communicate things like economic status, career status, social status, age, and personality. People use these cues, whether they are aware of it or not, in interpreting what you say.

The major guidelines for dressing to teach will be common sense and the culture at the place where you are teaching. If you have any doubts, you will be safer leaning toward the formal and conservative—in other words, typical "professional" dress.

Pay some attention to what colors and styles you look best in; you want your clothes, hair style, jewelry, etc., to be flattering but not distracting.

### Body Language

If your clothes say a lot, your body says even more! Where your general appearance says much about who you are, your body says a lot about how you feel, right now. Your facial expression, your posture, your body movement, and your gestures all communicate such things as your attitude, your mood, and your emotions.

As you stand in front of the classroom, no matter what you are wearing, an observer can immediately determine how you feel about yourself today. A number of physical signals can betray nervousness: trembling hands; perspiration; minimal movement; or nonstop movement; red face; a worried look.

Here are some typical body messages that you may want to practice.

> Posture—a posture that is erect without being stiff can communicate confidence and ease. If you keep your back straight, your shoulders relaxed and not hunched, your arms hanging naturally at

your sides, and your weight evenly distributed on both feet, you will appear comfortable and relaxed.

*Movement*—since movement helps keep the audience alert and provides some variety, you want to be comfortable moving as you teach (see "Getting Physical"). Smooth, natural strides, as opposed to tight or jerky steps, will communicate poise and comfort. If you appear ill at ease, your audience will be uncomfortable, too.

*Gestures*—use gestures naturally to punctuate your speech, but try not to overdo them. The best way to know what is too much is to ask for feedback, preferably in a walkthrough prior to the real presentation. Arm and hand gestures that are somewhat "bigger than life"—that is, large and away from the body—are good for emphasis. Gestures that keep your hands and arms close to the body may communicate apprehension or nervousness, as if you know that a gesture is appropriate, but you are afraid to really go for it.

*Facial Expressions*—let your whole face, not just your mouth, move naturally as you speak. If your face shows no change of expression at all, you may be perceived as boring or at least lacking in enthusiasm. You don't want your expressions to be distracting, but some variety appropriate to what you are saying at the time will make you and the presentation seem more interesting. Practicing in front of a mirror will help.

### Eye contact

Eye contact is perhaps the most important of the visual cues when you are communicating. If you cannot even look at the person you are speaking to, it doesn't much matter what you are saying: the message will be the interpretation of the eyes.

Eye contact communicates things like interest and involvement, confidence, honesty, respect and trust. Lack of eye contact can communicate fear, nervousness, lack of confidence, lack of respect, and untrustworthiness.

As an instructor, you will have more than one person to maintain eye contact with, so a walkthrough is a good time to practice and get feedback about how you come across.

Ideally, you want to make eye contact with as many people as you can, but not all at once. The usual recommendation is that you maintain eye contact with one individual for about five seconds before making contact with another.

These are some typical pitfalls to practice avoiding:

Looking at only one person most of the time.

Looking at only one side of the room most of the time.

Scanning too quickly so that it seems you never quite have eye contact with anyone.

Never quite looking directly at the audience—this is by far the worst offense.

### The Conceptual Message

When people interpret what you say, their interpretations are based on not only what they hear and see but also on their conceptual image of you:

Who you are (sex, race, nationality, age).

What you do (professionally, personally).

Where you come from (professionally, geographically).

Your knowledge/experience (professional, personal).

Your status (professional, social, economic).

Some of these details will be obvious when people see or hear you, and others will not. Participants may come into class already knowing something about you, but as you communicate with them, the verbal, vocal, and visual messages will add to their conceptual image.

It will be up to you what information you share and how you shape your image in the minds of others. This is one reason your introduction at the beginning of the class is so important. This is your chance to tell and show the audience what you want them to know about you. (See "Getting the Class Off to a Good Start.")

If you are afraid that the conceptual image your audience has of you may work against you—for instance, if they were to know all the experience you *don't* have—then work on playing up the positive angles that will work for you. You needn't point out your weaknesses, but you do want to mention the relevant strengths.

You may have less control over what you communicate conceptually than over what you communicate in other verbal and nonverbal ways, but being aware is a first step. It also helps to be aware of your audience's preferences: perhaps they consider outside vendor instructors more credible than in-house instructors —or the other way around.

You don't have to buy into preferences that don't match who or what you are. What you believe about yourself is what you will project and what others will perceive.

## *CHANGING YOUR SPEAKING PATTERNS*

The last few sections have highlighted aspects of verbal and non-verbal communication and have encouraged you to evaluate your own style for effectiveness.

Here are some general tips for those who have identified aspects of their speaking styles that they would like to change: specific recommendations for things to practice are included in the relevant sections.

**1.** *Hear yourself.* Hear your patterns, and decide what you want to change. A tape recorder is a wonderful aid here: you can tape record part of your presentation, or you can tape record yourself in casual conversations (such as your end of a phone conversation).

**2.** *See yourself.* If possible, videotape your presentation so that you can observe the visual message you send.

**3.** *Get feedback* from others, especially on the visual and conceptual characteristics, since you can't tape record those. One way to get feedback is to have a walkthrough of your presentation and let your audience know specifically what feedback you are looking for. Also, you might make a game out of having a friend catch you every time you fall short of your verbal or nonverbal goals.

**4.** *PRACTICE!* Keep practicing and then record yourself again, or have another walkthrough and notice the changes.

**5.** *Avoid being hard on yourself.* Know that it is OK to find things in yourself you want to change; and take the energy you would have spent on self-criticism and spend it instead on making the changes you wish to make.

## *HOW YOU LISTEN*

A big part of being an effective communicator is being a good listener. As an instructor, you must do more than simply deliver information; you must also effectively receive what your audience communicates to you.

Unfortunately, in the same way that many technical instructors have never been formally trained in communication skills, many who have been trained as speakers have never been trained as listeners.

Here are some tips that can help you become a better listener in your role as a teacher.

**1.** Have an *open attitude*. Assume that anything a participant has to ask or say is important.

**2.** *Let your interest show.* Maintain eye contact with the person speaking, and allow your face to reflect interest rather than boredom, impatience, or judgment.

**3.** *Be responsive.* Nod or shake your head in response; say things like "Mm, hmm," to let the speaker know you are following what is being said; acknowledge the contribution appropriately (answer the question, comment on the comment or answer, etc.) before moving on to something else.

**4.** *Don't create or tolerate disruptions* while someone is speaking to you. This may mean asking others in the class to be quiet while a participant is talking.

**5.** *Allow the speaker to finish* unless you need to interrupt to regain control of the discussion or the flow of the class.

**6.** *Hear the whole message* before responding. Don't assume that you know how someone will finish a sentence or a thought; you may distract yourself by preparing a response to something that is *not* being said and then miss what really is being said.

**7.** *Be an objective listener* as much as possible. This means being an egoless listener in not taking comments personally unless they are clearly meant that way. If you are teaching Word Perfect and someone declares an unflinching preference for DW4, you want to avoid taking it personally and reacting emotionally. That kind of response can damage your credibility, and discourage that individual and others in the class from contributing any more questions or comments.

**8.** *Don't let emotional words block the message for you.* We all have key words or subjects that can trigger an emotional reaction (push our buttons). If we allow these triggers to go off while we are listening, we run the risk of responding inappropriately and not hearing all that is being said.

**9.** *Be a critical listener.* Look for the real meaning of what is being asked or said; ignore things like the speaker's accent or his subjective asides, which can distract you from the real content of his message.

**10.** *Be nonjudgmental.* Mentally critiquing the speaker or her delivery will distract you from hearing the content of the message.

# *The Presentation*

## *PURPOSE OF THE PRESENTATION*

Whatever you are teaching, the ultimate learning objectives fall into one of three basic categories: cognitive, psychomotor, and affective. Effective teaching means varying your techniques depending on the type of learning you are causing to occur.

Below are examples of each type of learning, along with related teaching techniques.

### *Cognitive*

The learning objective is to understand, to be able to explain.

#### *Examples*
Participants will be able to explain the rationale of putting an application on the mainframe instead of on a PC.
Learners will be able to describe the features of a typical mainframe database management system.

#### *Teaching*
Use lecture with visual aids.
Use guided discussion where appropriate.
Use clear, realistic examples.
Discuss case studies, if appropriate.

*Psychomotor*

The learning objective is to demonstrate a physical or mental skill.

### Examples

Learners will be able to code all sections of a COBOL program.
Participants will be able to give a technical lecture.
Participants will be able to use a keyboard.

### Teaching

Use lecture with visual aids.
Use guided discussion, where appropriate.
Use clear, realistic examples.
Provide as much hands-on practice as possible.
Provide feedback on learners' progress.


*Affective*

The learning objective is to experience a change in attitude as a result of the class.

### Examples

Learners will demonstrate enthusiasm for using structured systems development methods.
Participants will be convinced that using code-generators is the only way to go.

### Teaching

Use dynamic lecture.
Use impressive visual aids.
Use a lot of guided discussion, giving participants the opportunity to express their feelings and ideas.
Demonstrate the qualities you want the audience to emulate. For instance, if you want your audience to be enthusiastic, *you* must be enthusiastic.


## PRESENTATION METHODS

As a stand-up instructor, you can present information to a class in a variety of ways. Each of the formats mentioned below will be effective in different situations, depending on the purpose of the class and the learning objectives.

However, that doesn't mean that only one method should be used throughout an entire class. Most classes lend themselves to more than one method, and, in fact, you will be a far more effective instructor if you use a mix of formats where appropriate. Switching methods provides variety for both you and the audience and is an effective way of extending the attention span of the participants.

The teaching methods most commonly used in technical classes are described below, along with some tips for using them effectively.

### *Lecture*

#### *Basic Format*

You present; your audience listens. There is little or no interaction between you and the participants, depending on whether they ask questions.

#### *When to Use*

Lecture is appropriate when the learning objectives are cognitive and affective, and can be used with other methods when teaching psychomotor skills. It is effective any time when what you are teaching is something your audience has no experience with. It is by far the most frequently used presentation method.

#### *How to Use*

In a technical learning situation, lecture is most effective when enhanced with visual aids; when there is plenty of question and answer time; and when it is broken up occasionally by guided discussion, exercises, workshops, and breaks.

#### *Advantages*

Lecture is a direct, concise way to impart conceptual information.

*Pitfalls*

No matter how interesting your topic is, straight lecture for more than 20 minutes challenges the attention span of most adults in a learning situation.

### Guided Discussion

*Basic Format*

You lead your audience in a discussion, the objective of which is to let participants develop or demonstrate the point(s) you are trying to get across. Participants may interact with you or with each other, or both.

*When to Use*

Guided discussion is most effective when your audience has some knowledge about the topic.

*How to Use*

A discussion may start with your asking a question or with your audience asking a question of you. As the discussion progresses, your answers and subsequent questions and comments lead the audience in the direction you want them to go. You control the interaction so that it does not get off track; it is not a free-for-all.

*Advantages*

The discussion helps participants integrate or recall what they already know and reinforces the learning. It is also a great way to get people involved—something you want to do with an adult audience.

*Pitfalls*

If you do not control the discussion, it can quickly get off track or take a lot more time than you had planned.

## Demonstration

### Basic Format

You demonstrate (as opposed to talk about) what it is you want your audience to know or do. In technical, computer-related classes, a demonstration is generally of a hardware or software product.

### When to Use

Use demonstration in place of lecture when you can. It is the preferred method for teaching software overviews and classes that teach skills in using software.

Another way to use demonstration would be in conjunction with a guided discussion to develop a data flow diagram, for example; you're getting input from the class as you develop the diagram on the board, and the activity is demonstrating the process.

### How to Use

Always plan a demo and practice it before class if you possibly can. There may be times when you have an unforseen opportunity to demonstrate something, in which case think quickly on your feet and make the demo as straightforward and organized as you can. See "Conducting Demonstrations," for tips.

### Advantages

Seeing something work is much more effective for the learner than merely hearing about it, even with great visual aids.

### Pitfalls

A demo that is not well-planned and well-organized may be worse than none at all. It is important that participants be able to clearly see and understand what you are doing.

## *FEEDBACK/EVALUATION APPROACHES*

Part of your job as an instructor is to periodically evaluate whether your audience is with you. Do they understand? Do they have questions? Is the expected learning taking place? Do they need a break? Do they need help? If so, where?

To evaluate these things, you need feedback from the participants. Here are some suggested ways to obtain feedback, some of which you will want to plan for, others of which will be more impromptu.

### *Encourage Questions and Comments*

Encourage participants to ask questions and make comments. This way, they can let you know directly what is on their minds.

> Mention during your introduction that people may feel free to ask questions.
> Pause after making major points to give people time to assimilate what they've heard and think about whether they have questions.
> Ask "Are there any questions?" at appropriate points, such as at the end of topics and right before breaks or at the end of the day. Then pause long enough that people have time to think.

### *Take A Fog Count*

Taking a "fog count" means reading your audience visually and intuitively to pick up problems or questions that are not being expressed overtly. Even when you encourage interaction, there will always be some people who do not speak up: they may feel self-conscious; they may think their questions or comments will not interest the group; or they may have a question, but just not know exactly what it is.

To "hear what is not being said," you want to develop your skill at being able to read facial expressions and body language. Generally, participants who sit up straight, maintain eye contact, and respond with nods are not the ones you need to worry about. Blank looks, lack of eye contact, slouched posture, audible groans, and general lack of participation are warning signals.

Learners can nonverbally communicate such things as:

I'm bored—I wish you would pick up the pace.
I'm confused—I wish you would slow down.
I'm confused—would you repeat that last point?
I'm totally lost—you're wasting your time to continue.
I've got it! I understand!
I'd like to share an experience.
I have a question, but I'm not sure yet what it is.
I need a break.
It's too late in the day for me to absorb anything else!

If you're not sure what your audience is communicating, you can always ask.

### Ask

Of course, being perceptive is only half the job; the other half is responding appropriately to what you pick up. The first thing you want to do is verify what you are perceiving, and generally you will do this with a direct question:

"Are you with me?"
"Shall I repeat that?"
"Joe, you look confused. Do you have a question?"
"Where did I lose you?"
"Are you ready for a break?"

### Give a Review

A review can be oral or written, and it is your opportunity to get feedback directly—by asking questions designed to tell you whether your audience understands the points you think are important.

Make sure your questions—whether written or oral—are asking things you really want to know and are designed to encourage learners to think and to integrate what they have learned or experienced in your class. For example, the answer to "What does JCL stand for?" will not tell you nearly so much as the answer to "When would you use an instream PROC as opposed to a catalogued procedure?"

A review for feedback can be combined with the kind of review in which you summarize important points at the end of a learning unit. This way, you can be sure the audience understands what you just taught before learning something new.

Reviews are also effective lead-ins at the beginning of class each day and after lunch. These are times when people, having been away from the classroom for an hour or overnight, may understandably have other things on their minds besides what you were talking about before the last break. Besides giving you feedback on what they know, these reviews also help learners focus again on what they are learning and get their minds off the traffic, the late babysitter, the angry boss, or the attractive lunch date. And while they are making that transition, they haven't missed anything new.

### Assign Written Exercises

Written exercises are really a form of review, giving learners the opportunity to practice repeating things they have just learned. For example, matching and fill-in-the-blank exercises are especially effective for reviewing cognitive learning, where the objective is for participants to gain knowledge and understanding of concepts and ideas.

Exercises can also be a way to practice organizing steps or ideas that will later be used in a hands-on workshop. For example, you might have participants write out their COBOL code in a seat exercise so that you can review it with them before they enter it at a terminal and compile it.

### Conduct a Hands-on Workshop

Observing learners in a hands-on practice situation is a great way to see how well they have mastered the skills you are teaching, and to find out what questions they have. Even with plenty of chances to ask questions during lecture, most learners won't really know what questions they have until they actually start doing what they've been hearing about.

Workshops also give you an opportunity to coach one-on-one and reinforce what you've taught.

# Teaching Aids

## WHY YOU WANT TO USE TEACHING AIDS

If you have ever attended a class in which all the information was delivered verbally by the instructor—with no pictures, no drawing on the board, no handouts—then you can surely appreciate the value of teaching aids.

A teaching aid, as the name implies, helps you teach. It can:

Help you deliver information.
Help you clarify or emphasize a point.
Make it easier for learners to follow along.

Besides making your job easier, teaching aids provide other benefits as well.

**1.** They *increase retention* by reinforcing the spoken words of your lecture.

**2.** They *provide variety*, which helps your audience stay alert and interested.

**3.** They *save time*—as they say, a picture is worth a thousand words.

One of the easiest ways you can make your class more interesting is to use teaching aids. And remember, you are teaching adults, who can choose to leave if they are not interested.

## TYPES OF AIDS

Usually, when we think of teaching aids, what comes to mind is visual aids; but teaching aids can also include handouts. The following chart describes the most frequently used types of teaching aids and lists their advantages.

## USING AIDS EFFECTIVELY

Here are some pointers to help you in using the most common teaching aids.

| Type | Advantages/disadvantages |
|------|--------------------------|
| *Transparencies* (foils): these are usually prepared pictures or bullet points, but blank ones may be used to write on as you go. | ( + ) Can be written on to add to what is already there.<br>( + ) Can easily be seen by everyone.<br>( + ) Can be used without dimming the lights<br>( + ) Can serve as a prompt to the instructor.<br>( + ) Can be revealed one point at a time.<br>( − ) Must be kept updated. |
| *Flip charts:* can be prepared or written on as you go. | ( + ) Pages may be saved to refer to again later.<br>( + ) Pages may be hung around the classroom.<br>( − ) Charts are not as easy to see as foils.<br>( − ) They don't give you much room to write, compared to a board.<br>( − ) Revealing one point at a time is cumbersome. |

| Type | Advantages/disadvantages |
|---|---|
| *Board:* either chalkboard or marker board. | (+) Boards usually give you lots of room to write, giving you space to add on as you go. <br><br> (+) Examples may be left on the board for people to refer to while other aids, such as foils, are being used to develop other points. <br><br> (−) Chalk can be a nuisance to work with. |
| *Screen projection:* the video projection of a computer or terminal screen. | (+) This is a great way to demonstrate an on-line system or the use of a software product. (See "Conducting Demonstrations.") <br><br> (−) You need special equipment, which can get very expensive. <br><br> (−) You must usually dim the lights to get a clearly visible picture (people may doze off). |
| *Films:* either 16mm or video cassette. | (+) Films provide a pleasant change of pace if they are short and relevant. <br><br> (−) You need special equipment, which may not already be in the classroom. <br><br> (−) You must usually dim the lights. |

| Type | Advantages/disadvantages |
|---|---|
| *Slides* | Same as films |
| *Handouts:* printed copies of information that you either give participants at the beginning of the class, or hand out at appropriate points during the class. | ( + ) Handouts that are well-designed and organized make it easy for participants to follow along.<br><br>( + ) They give participants a handy place to take notes.<br><br>( + ) They give participants something to refer to after class.<br><br>( − ) Preparing handouts can be very time-consuming.<br><br>( − ) Handouts must be kept updated. |

### *Transparencies*

**1.** Have your projector lined up and focused before showing your first foil.

**2.** Be sure your foils are clean and in order before you begin.

**3.** If your lecture no longer relates to the foil that is showing, remove it and turn off the projector until you are ready to use another foil.

**4.** To point to something on a foil, use a pointer and point to the projection of the image; or lay the pointer or a pen over the foil so that its shadow is projected with the foil.

**5.** Use water-soluable colored pens to highlight points on the foils as you lecture. You may want to add some permanent color in places to make the foils more interesting (most foils are either black on clear or black on a colored transparency; generally, multicolored foils are too expensive and time-consuming to use in the volume you may need them in a technical class.

**6.** If your foil has several bullet points, and you want to talk about each one briefly before going on to the next, you can use a sheet of paper to reveal one point at a time.

**7.** Try to keep the foils in order as you use them, so it will be easier to find one you may want to pull out and show again.

**8.** If your foils have cardboard frames, write your prompting notes on the frames. They will be easy to read without being obvious.

**9.** Get into the habit of cleaning and organizing your foils right after class. This way, they will be ready to go the next time.

### Flip Charts

**1.** If you are preparing prewritten charts (as opposed to writing on them as you go), use every other page. If there is something written directly under the page you are showing, it will show through. At worst, it will obscure the page you are showing; at best it will distract your audience.

**2.** When you write on a chart, make your words large, brief, and easy to see. Use color to highlight points and make your charts more interesting.

**3.** To simplify your task of turning every other page, use paper clips to clip each page you have written on to the blank page underneath it.

**4.** If you need them, write prompting notes in light pencil on the edges of your charts. Put them at eye level so you can see them easily —they will be invisible to your audience.

**5.** Tear off charts you are finished with after the class. This way, the blank pages between charts can be freed up for use if you don't need the pages that were written on.

### Board

**1.** Write large.

**2.** Erase what no longer relates to what you are lecturing about.

**3.** If you have sliding boards that cover each other, you may want to write on one board ahead of time and have it ready to reveal at the appropriate time.

### Films/Slides

**1.** Have equipment set up and ready ahead of time.

**2.** Rehearse to check out the best lighting and the best position for visibility.

**3.** Remember to rewind films or cassettes.

*Handouts*

**1.** Have all handouts copied and ready to hand out before the class begins.

**2.** A rule of thumb is, don't distribute handouts until you are ready to talk about them. In technical classes, it has become customary to give participants a copy of all the visual aids and maybe some student notes at the beginning of the class, so they may follow along and take notes. Any other handouts should wait until the moment you are ready to refer to them, because people will be distracted by wanting to look at them the minute they see them.

*General Tips*

**1.** Make sure any visual is simple and clear; it should truly be an aid, not a distraction or a hindrance to communication.

> Use large letters.
> Use few words. A standard guideline is to limit bullet points to 6 or 7 and limit the number of words on one visual aid to about 20.
> Let each chart/foil represent a single idea.

**2.** Always rehearse the use of your aids, especially if you are using more than one at a time (like using both foils and the board or a flip chart).

> Know the best time to use them.
> Use them in the proper sequence.
> Make sure you know how to use any special equipment.
> Insure visibility and audibility.
> Make sure you have enough room to maneuver.

**3.** Avoid turning your back to the audience while you are using an aid. If you are writing on a flip chart or board, stop talking while you are writing.

# Presentation Tips and Techniques

## DEALING WITH NERVOUSNESS

It is normal to be nervous before giving a presentation or teaching a class, whether it is your first time teaching, your first time teaching this course, or your first time with this audience. If that seems to cover every situation, you've got the idea: it's just normal to be nervous before putting yourself on the line in front of a group of people.

Of course, feeling nervous can be uncomfortable, but being a little bit nervous can work to your advantage in helping you remember to prepare.

Why do we get nervous? Usually out of fear. Here is a list of typical things we as instructors fear and some things you might do to overcome those fears.

We fear we are not qualified to be teaching the class. There is some reason that it is you rather than someone in your audience who is teaching this class. Odds are you are teaching because you know what it is that your audience needs to know, even if they are more knowledgeable than you in other areas. Maybe they are all vice presidents, or maybe they have all been technical many years longer than you have. But they are coming to this class because they DON'T know what you know about this subject.

Remind yourself of the reasons you are the one teaching. Concentrate on the things you DO know, that you CAN do, rather than on your real or perceived shortcomings.

Verbally affirm to yourself the attributes that you want to see yourself possessing. For example, as you are showering or driving to work, you can tell yourself "I am thoroughly knowledgeable about (whatever)"; "I am confident"; "I am the best person to be teaching this class"; and so on. Our beliefs become self-fulfilling prophecies, so begin convincing yourself that you are qualified and will be successful.

We fear we won't give a smooth presentation. Nobody likes to make mistakes, and it can be especially uncomfortable to make mistakes in front of an audience.

Practice is the best way to overcome this fear. The more you practice the skills you want to improve, the more comfortable and less vulnerable you will feel.

If you are worried about your speaking style, practice with a tape recorder. (See "Communication Skills.")

If you are worried about things like facial expressions, body language, or what to do with your hands, practice in front of a mirror. (See "The Visual Message.")

If you are afraid you will forget what you want to say, prepare instructor notes you can refer to while you are teaching.

If you are worried that the topics will not flow well, walk through your presentation with an audience and get feedback. A walk-through, in any case, is excellent preparation. (See "Walk-throughs" and "Controlling Class Flow.")

Practice mentally by using creative visualization. Relax with your eyes closed and picture yourself going through all or part of your presentation. Focus on the areas where you anticipate having a problem and "see" yourself performing the way you would like to. This is an extremely effective way to program yourself for success. (See "Visualizing Success.")

We fear we will finish too soon or run too long. Especially when we are teaching a class for the first time, we may not have a clear sense of the time we will need to cover everything.

Again, walk through as much of your presentation as possible, if not physically, at least mentally. Try to anticipate questions that will come up and require discussion, practice all exercises and workshops, and practice your guided discussions.

**We fear we will not be able to answer questions.** We assume that if we don't know all the answers, we are not qualified to teach. Or we think our technical integrity will be in question.

> First, it helps to understand that there is nothing wrong with not knowing an answer occasionally. (See "When Your Answer is 'I Don't Know'.")
> Second, if you establish your credibility at the beginning, you'll find your audience is usually very forgiving. (See "Getting the Class off to a Good Start.") Most people understand that nobody knows all the answers all the time.
> Third, if you truly cannot answer *many* questions, learn more about your subject matter—*before* you teach. (See "Know Your Subject.")

**We fear the audience will be bored.** Images of our own least favorite instructor come to mind, and we dread the thought that participants in our classes might be as bored as we have been.

> Think of the things that make you bored in a class. Usually, if you are bored, it is because the pace is wrong, you already know the material, or the instructor simply does not hold your interest. When you practice, avoid the things that you know cause you to lose interest.
> Learn to read your audience's moods. Skill in reading your audience will give you confidence that you will pick up early if people are bored and be able to remedy the situation. If you are not sure, ask for feedback from the class. (See "Feedback/Evaluation Approaches.")
> *If your audience really is bored, remedy the situation:*
> Change the pace.
> Involve the class in an interactive discussion.
> Change your teaching method—switch from lecture to an exercise, workshop or a demonstration.
> Take a break.
> Use humor to break the flow.

Feeling nervous is more that just mentally worrying about how the class is going to go. Nervousness manifests itself physically as well. Following the tips above should help you overcome the mental stress, which will surely alleviate many of the physical symptoms. But sometimes your body will act nervously anyway. When you're nervous, you may experience some of these symptoms:

a queasy stomach
a dry mouth
shaky knees
a quivering voice
perspiration
giggling
faster speech

Here are some techniques that may help you physically relax.

Take several deep breaths. Inhale and exhale slowly, allowing your whole body to relax.

Bite your tongue gently to produce saliva. Better to try this one before everyone's eyes are on you.

Meditate right before class. If you are a regular meditator, you know the relaxing benefits of this practice. If you have never meditated, try sitting quietly in a comfortable position with your eyes closed; take several slow, deep breaths, and focus your mind on something like a word or an image or a sound (the sound of your breathing, music); try to keep random thoughts from taking over—let your mind relax, and your body will begin to follow suit.

Smile. A smile requires relaxed facial muscles, and even if you don't have time to do anything else to help yourself relax before a class, you can walk in with a smile. A relaxed face is a start toward a relaxed body. Try it, and notice the effect that simple act has on both you and your audience.

Laugh. If you are comfortable starting out with some humor, you can get yourself and the whole class laughing, and that *really* helps everyone relax. Just be sure the class is laughing with you and not at you!

Finally, remember that feeling nervous can work in your favor by reminding you to focus on what you are doing and to be prepared. Instead of being hard on yourself every time you feel nervous before a class, try accepting those feelings as part of your professional edge, and let them work for you rather than against you.

## GETTING THE CLASS OFF TO A GOOD START

The way you begin your class will set the mood for everything that follows. In fact, probably the most important few minutes in any class are those at the very beginning. The way your audience interprets what you say and do at the start of the class will largely influence

The learning environment you create
Your credibility with the audience
Your control over the flow of the class

Besides the fact that you want to have credibility, control, and an effective learning environment, there is another advantage to getting off to a positive start: if anything goes wrong later in the class, your audience is much more likely to be forgiving if you've gained respect and established credibility in the beginning. If you start out making mistakes, you are playing catch-up, maybe for the entire class.

Since how you start your class is related not only to what you do standing in front of your audience but also to things you've done beforehand, the following suggestions are grouped accordingly.

### Before Class

**1.** Be psychologically ready to teach. Relax and know that you will do well. Regardless of what else is on your mind, focus now on your class. Try closing your eyes and visualizing your class going exactly the way you would like it to go. (See "Visualizing Success.")

**2.** Be physically ready to teach. Be well-rested and well-nourished. No matter how last-minute your preparation, remember to allow yourself a good night's sleep before a teaching day. And you may find that you will have more energy when you start out the day with a nourishing breakfast and are more alert in the afternoon if you have a light lunch.

**3.** Know your audience. Know who they are and what they are expecting. (See "Know Your Audience.")

**4.** Have the classroom set up the way you want it before the participants arrive. (See "Preparation Checklist.")

**5.** Be organized. Have ready all your teaching materials, handouts, visual aids.

**6.** Be ready to start on time.

**7.** Know exactly how you will begin. Even if it is a class you have taught many times, rehearsing your opening can give you added self-confidence during those crucial opening moments.


### In the Classroom

**1.** Create a friendly atmosphere. Since you are prepared and organized, you can relax and greet people as they come in.

Smile—especially if you don't have time to do anything else. A smile is by nature relaxing, and when you smile, people generally smile back—it helps everyone relax.

**2.** Introduce yourself. Your audience will want to know not only your name but also your background relevant to the subject matter of the course. This is where you can establish your credibility by mentioning your experience. You can bolster your credibility by promoting your strengths and by not advertising your weaknesses, not making excuses, and not blaming others for your shortcomings or those of the course. The following suggestions are valid not only during your introduction but at any time during the class.

Mention your strengths.

*"I've been using this software for about two years now, and in my last job, I was responsible for training all our new people in how to use it."*

Avoid mentioning your weaknesses.
Instead of

*"I've never taught before,"* or

*"I've been teaching this course for several months, but I've never really worked with this software,"*

Try

*"This course covers a wide range of topics, and I don't claim to be an expert in all of them."*

That's a great way to avoid saying that you are not yet an expert in *any* of the topics.

Don't apologize for yourself or the course; don't put yourself down.

*"I don't draw very well, but . . ."* as you are putting an illustration on the board.

*"I'm sorry these foils are handwritten, but . . ."*

Don't make excuses.

*" . . . but our support person has been on vacation this month."*

*"I just found out yesterday they wanted me to teach this class."*

*"The stock market crashed yesterday, so I haven't finished checking your programs."*

When it is appropriate to apoligize, simply apologize and then drop it.

*"I apologize for being late. Let's get started."*

Use humor if it is appropriate to the situation and if it is a comfortable part of your natural style.

**3.** Have participants introduce themselves. Have them tell you things that will assist you in being an effective instructor.

Find out what experience, if any, they already have with the subject you are teaching. You'll be able to tell who are the most and the least knowledgeable, and this will help you know what level to teach to; how you may want to team people up during workshops; who might need special help; and who might need an extra challenge to avoid boredom.

Find out what each person hopes to get out of the class. Not only will this information help you tailor your presentation, but it helps you and the participants determine up front whether anyone is in the wrong class.

You can establish good rapport during these introductions by maintaining good eye contact and showing genuine interest in your participants and their reasons for being in class.

**4.** Give an overview of what the audience can expect.

Learning objectives.
Schedule—when you'll begin and end each day, when you will go to lunch and take breaks.
Format (lecture, workshop, etc.).
Scope of course—what it does and does *not* cover.
Materials—what participants should have, and what they may keep.
What will be required on their part—attendance, homework, assignments that have to be turned in, etc.

**5.** Cover miscellaneous details.

How can people reach you when you're not in class?
Where are the restrooms?
Where is an available phone?
Where can people pick up messages?

Remember, these are only suggestions. As you teach more and more classes, you will discover what works best for you, and you will probably develop your own routine opening. Also, remember that each class is different, and your opening should always be appropriate to the particular audience you are addressing.

## *CONTROLLING CLASS FLOW*

Part of your success as an instructor will depend on how well you control the flow of your presentations. Controlling the flow generally means these things:

The class begins *when you want it to.*
The class ends *when you want it to.*
Class resumes after breaks *when you say it will.*
Guided discussions stay *on track.*
You cover all the topics *you intend to cover.*
You can entertain side discussions *without going beyond the scope* of the course.
If any problems arise, you handle them *calmly and decisively.*
There is never any doubt that *you are the one in charge* of the class.

Naturally, you want to control the class in a way that commands the respect of your audience without making you seem like an obnoxious bully at the same time. Here are some suggestions to help you accomplish that goal.

**1.** Be punctual. If you are in the habit of starting on time, stragglers will soon get the message and be on time themselves.

Begin the class on time.
Take breaks when you say you will.
Start up after breaks when you say you will.
End the class on time.

**2.** Communicate control with your speaking style. Do not be afraid to be assertive. (See "Communication Skills.")

*Speak with Courteous Authority.*

Avoid a high-pitched, very fast or very soft voice.

When making a statement, especially an emphatic one, avoid ending with an upward pitch as if you are asking a question; that is, bring the pitch down at the end of the sentence.

Avoid meaningless phrases that withdraw your authority, like "OK?."

Use pauses and silence for effect. Don't feel you must fill every quiet spot with chatter.

*Be Assertive.*

Speak with confidence when recovering from a mistake or admitting you don't know an answer.

Use confident body language: maintain good eye contact with your audience; stand up straight; avoid nervous habits, such as wringing your hands.

Politely take control of sidetracked discussions:

*"That's a good question, but since it's beyond the scope of the course, let's discuss it after class.*

*"Those are all good points, and I'm glad you brought them up; getting back to the point I was making . . ."*

Be firm with problem participants. (See "Dealing With Problem Participants.")

**3.** Know your course. It is hard to control the flow of the course if you don't know how it should flow in the first place. (See "Know Your Course.")

Know the order in which you will present topics.

Know the timing required—or make an educated estimate—for lectures, workshops, and guided discussions.

Know the scope of the course, so you know not only what should be covered but also what should not be covered.

Practice with a walkthrough to get a sense of timing, anticipate questions that will arise, and test the logical flow.

If possible, sit in on the class when someone else is teaching it.

**4.** Handle interruptions confidently and courteously. Don't be afraid to exert your authority when necessary to maintain control of the class. Do remember that you are dealing with adults. Usually a polite reminder will be enough to give participants your message.

One good way to prevent certain predictable interruptions is to mention during your introduction what you expect:

*"We'll have plenty of time for the hands-on workshops, so I'll expect you to refrain from using the terminals/PCs during the lecture portion of the class."*

Then if the situation does arise later, your reminder at that time will not be a surprise; nor is it so likely to put the offender on the spot:

*"When you've all signed off, I'll begin."*

*"Everybody please stop keying now. You'll have plenty of time to finish the exercise in the next workshop."*

Always remember that you are in charge and, uncomfortable as it may be at times, you have a right to assert your authority in order to stay on schedule and to control distractions or interruptions. Remember, too, that no matter how annoyed or frustrated you may get, you want to always treat your audience—collectively and individually—with respect.

## MAKING A TRANSITION TO THE NEXT TOPIC

An important ingredient in a smoothly flowing class is effective transition from one topic to the next. You may know the course thoroughly and be right on track with your timing, but if you are losing your audience to confusion as you switch topics, you have sacrificed some of your effectiveness.

A good rule of thumb with any kind of presentation is to follow this advice:

Tell 'em what you're gonna tell 'em.
Tell 'em.
Tell 'em what you told 'em.

While this approach is recommended for a presentation as a whole, it is also appropriate for individual topics within a presentation or class. Transition points then become a combination of "tell 'em what you just told 'em" and "tell 'em what you're about to tell 'em next." But a good transition does more than simply wrap up one topic and introduce another; it also explains the relationship between the two topics.

With that in mind, here is a suggested approach to the transitions you need to make.

**1.** Wrap up the topic you just finished.

Summarize—tell 'em what you told 'em.
Review—This can be a combination summary and feedback session; find out whether objectives for the topic just ended have been met.
Are there any questions? Make sure there are no questions outstanding before you continue to the next topic.

**2.** Relate last topic to next topic. It helps an audience to know whether the next topic is related to the last, and if so, how.

If you are teaching a class of several PC tools, one topic might deal with spreadsheets, and the next might be word processing.

*"Now that you are comfortable with spreadsheets, we will learn how to do word processing. We'll be working with completely different software, although later we will see how you might use the two together."*

If you are teaching a database design class, your transition might be from talking about the logical database constructs to presenting the physical design options.

*"We've just examined the logical structure of a network database. Now we'll look at how we can physically implement that logical structure. Again, the logical design must be complete before we can tell the system how to physically implement that design."*

With some classes, it helps to refer back to an outline or overview diagram and point out that "you are here" for reference. For instance, if you are teaching a project management class, you might be teaching several topics related to each phase of project development. As part of your transition, you will relate the previous and next topics, and you might also want to point back to a handout showing the different phases and remind everyone that you are still within the planning phase, say, or switching now from this phase to that phase.

**3.** Introduce the next topic. Tell 'em what you're going to tell 'em.

Remember that a good transition makes it clear where you've been, where you're going, and how the two relate.

## *EFFECTIVELY TEACHING TECHNICAL MATERIAL*

Being an effective technical instructor requires more than subject matter expertise and good presentation skills. It also helps to have an awareness of the fact that technical topics are often perceived as dry and boring. If this is the perception your audience has, chances are they are not going to learn as much or have as much fun learning as they would if they found the subject interesting.

The perception "dry and boring" is a judgment on the part of the learner, usually for one of two reasons: the learner is not interested in the topic, or the instructor has not held the learner's interest while presenting the topic.

Here are some suggestions to guide you in doing your part to help people learn technical material effectively.

**1.** Keep your audience alert by varying your presentation method. Lecturing for more than 20 to 30 minutes at a time can tax anyone's attention span, no matter how interesting the topic. You have to be the judge, by reading your audience, of how long you can lecture without losing their attention. Remember, if they are tired of sitting and being talked at, they will tune out and think about something else; you'll be wasting your time and theirs. Here are some alternatives to lecture you might try.

*Guided discussion:* Some technical topics don't seem to lend themselves to discussion, but you can usually at least ask some open-ended questions that get the audience involved in the learning process.

For example:

"Why would you use a PERFORM instead of a CALL?"

"What's wrong with the example on page 25?"

"Give me the statements we need to code the Environment Division, and I'll write them on the board."

*Exercises and workshops:* You will want these to be part of your course anyway, because they will reinforce the skills you want the participants to learn. The point here is that if you pay attention to where you include them, they are a great way to break up the lecture and keep people interested.

*Demonstration:* If you are teaching a topic you can demonstrate— and you have the equipment to do it—you can liven up a presentation by demonstrating how to do something, rather than simply talking about it or illustrating it with transparencies.

*Role Play:* When it fits what you are teaching, role play, like exercises and workshops, can effectively break up a lecture and, at the same time, get your audience involved in the learning process.

*Note: See "Presentation Methods" for more information about when and how to use these various teaching methods.*

**2.** Avoid overwhelming with facts. People need to assimilate what they are hearing. They need to have time to relate the new information to things they already know. A nonstop lecture can overwhelm learners with details without giving them time to know whether they have questions.

By using examples, analogies, and illustrations, you help learners grasp concepts more easily while you are also breaking up the straight lecture.

*Examples:* Use realistic examples whenever you can to clarify the details you are teaching. For instance, after lecturing about JCL syntax, show the class a listing of real JCL used in your shop. Not only will they see how real JCL statements are likely to be coded, but they will see how all the statements fit together.

*Analogies:* New technical concepts can be elusive at first. Effective analogies that describe the subject in terms people can relate to will help them understand faster. For example, saying that developing a computer system is kind of like building a house—something most people can easily relate to—and then describing each stage of development in terms of corresponding stages of building a house, can be very effective in helping a new analyst understand the nature and importance of each stage.

*Illustrations:* As they say, a picture is worth a thousand words, and given a choice, most technical audiences would surely prefer a picture to the thousand words. You can use predrawn diagrams on foils and flip charts, or you can draw, as the need arises, on a board or flip chart. You need not worry about artistic ability if you keep your drawings very simple. Besides, your audience is interested in grasping the concept, not critiquing your artistic talent.

**3.** Relate new information to your learners' reality. You are teaching concepts and skills that your learners will need to apply on the job. Be sure that in addition to the technical details, you also teach them to relate that information to their job situations.

Use realistic examples. Illustrate your points with examples from the environment in which the learners will be applying these new concepts/skills.

Allow hands-on practice. Actually doing what they are learning is probably the best way people learn. This way, when they get back on the job, they will already be comfortable using the new skills.

**4.** Be accurate in what you teach. With a lot of technical subjects, there are clearly right and wrong ways to do things. With more theoretical subjects, there may be no one right way. That being the case, you want to make sure that what you teach people is accurate.

Know the subject matter well before you teach it. If you are teaching a programming language and you can't explain how to avoid certain compiler errors, your audience will quickly lose confidence in you.

Have reference books available in the classroom. There is nothing wrong with looking things up—better to look it up than to give an incorrect answer.

If you don't know an answer, admit it and research it before the end of the class. People will not expect you to know everything, and your credibility will be stronger if you admit to not knowing than if you get caught guessing. (See "When Your Answer is I Don't Know.")

### ⟩ ⌐ *General Tips*

Probably the single most effective means of making a technical course really interesting is to get the audience involved. Remember that adults generally prefer to take an active role in their learning. As mentioned above, guided discussions, role play, and workshops are good ways to make your class interactive.

One of the hardest skills for new instructors to master, it seems, is the art of effectively and comfortably interacting with class participants in guided discussions. If you need some tips here, refer to the sections on "The Art of Asking Questions," "Responding to Answers," and "Responding to Questions."

## TEAM TEACHING

If you have ever had the opportunity to team teach a class, you know that success depends on more than just your own teaching expertise; successful team teaching also depends on your ability to coordinate effectively with one or more other instructors in a way that not only accomplishes your objectives but also gives your audience a smooth, flowing presentation. Your goal is to present your class in such a way that having two (or more) instructors truly enhances the learning experience.

### *Advantages*

When chosen for the right reasons, team teaching is an excellent presentation approach with definite advantages.

**1.** Added expertise—another instructor may bring to a topic expertise or experience that you lack.

**2.** Variety—more than one instructor can make a class more interesting by giving the audience a different face, voice, and style to focus on.

**3.** Assistance—you may need more than one instructor for things like workshops and role play.

*Possible Pitfalls*

When team teaching doesn't work, it is usually because of one of these reasons.

**1.** The class is confused about who is in charge. Is there a primary instructor with an assistant, or are you sharing the control equally?

**2.** The instructors compete with each other. They vie for attention, control, or credibility. No matter who "wins," all involved—instructors and participants—lose in this game.

**3.** The instructors' teaching styles do not complement each other. The two styles distract from the presentation more than they enhance it.

**4.** One instructor is perceived to have more credibility than the other. Participants end up directing their questions and responses to just one instructor, often asking the one to repeat or clarify what the other has said.

The tips that follow will help you avoid the pitfalls. They are mostly common sense, but they bear noting if for no other reason than to keep you from having to learn them the hard way.

*Before Class*

**1.** Decide roles ahead of time. Be very clear on who is going to do what. Make sure you and your teammate(s) agree on who is responsible for:

Preparation details—room, equipment, class materials.
Starting the class—who handles introductions, the overview, etc.
Teaching which topics—hopefully your respective strengths will balance.
Workshops—especially if you are conducting workshops together, be clear on your respective roles so that it is clear to the participants whether one or both of you are in charge.
Transitions—practice how you will turn over control to each other at each transition point. This is a good time to review what you have just covered, check and see whether there are any questions, and then briefly lead into the next topic; not only are you providing a clear transition from one topic to the next, but you are also making it clear that a different instructor is taking over.
If you are teaching with two or more other instructors, you may

find it less confusing for the audience if there is one instructor who does the introducing and transitions.

Ending the class—will one or both of you formally end the class?

**2.** Walk through the flow of the class. Practice how the class will flow with two of you teaching. Especially, practice your transitions from one speaker to the other.

### During Class

**1.** If you are not teaching, sit in the back of the room. Everyone will know you are one of the instructors, and if you are where the audience can see you, their focus may be divided between you and the instructor who is teaching.

**2.** Always treat your fellow instructor(s) with respect.

*Don't upstage the instructor who is teaching* by diverting attention to yourself.

*Don't disagree in front of the class*—resolve differences during a break.

*Respect the teaching instructor's wishes* with regard to whether you ask questions/offer answers while in the audience.

**3.** Provide clear transitions so the class knows who is in charge.

**4.** Present a unified image to the class. You are a team—you are working together, not against each other.

Team teaching at its best is very effective and fun for everyone involved. If you anticipate using this approach to teaching, it is worth taking some time to plan and practice. Good luck!

### DEALING WITH PROBLEM PARTICIPANTS

One typical fear, especially among new instructors, is that you will encounter a situation you can't handle. Not being able to answer a question or running out of time may make you really uneasy, but either of those situations may seem like a piece of cake compared to confronting a difficult participant.

Most of us are not accustomed to reprimanding or disciplining our peers, and having to do so in front of an audience may be extremely uncomfortable. But it is also very rare. Generally, adults in class behave like adults, which means they show respect for the instructor and for each other.

In the vast majority of cases, the problems participants present are the same problems you encounter in meetings where people interrupt, disagree, get off track, have a bad attitude, or tune out. In meetings, you may or may not be in charge; in the class, you are definitely in charge and accountable to a large degree for the success of the class.

Here are some general guidelines, followed by some specific scenarios and their suggested solutions.

**1.** Anticipate problems and prevent the ones you can before they occur. During your introduction of the class or of specific topics, you can tactfully let the class know what you expect with regard to anticipated problems.

For example, stating "I'll expect you to be using the PCs only during our workshop sessions and not while I am lecturing" accomplishes two things: it lets people know they are not to be keying while you are talking—a nice reminder for those who truly do not know any better; and it makes it easier for you later if you do have to ask someone to stop keying. Since you have already mentioned it, the person you are addressing will not feel quite so on the spot or singled out—you're not picking on her so much as just reiterating a rule you have already stated.

**2.** Be confident in your authority. Remember, you are in charge, and you have not only the right but the responsibility to resolve disruptions and keep the class flowing.

**3.** Always treat others with respect, as you would have them treat you.

**4.** Do not take another's behavior personally, unless it is clearly intended to be personal. If you take it personally, your emotions are likely to interfere with your effective handling of the situation.

**5.** Have a win-win attitude. Approach each situation with the attitude that you will mutually resolve it—not that there will be a winner and a loser. Helping the problem participant save face will do more for your integrity than will proving you are the "winner.".

**6.** Remember that you are a professional. Always assume that your dealings with a problem participant will not be limited to this class. Whether this individual is a fellow worker or a client from another company, assume that your paths will cross again. With that in mind, conduct yourself now so that future meetings will not be awkward.

## Typical Scenarios

The following scenarios illustrate some typical problems you may encounter and a recommended course of action.

### Who's Right?

#### Problem

Nancy Noitall strongly disagrees with a statement you have made. You've explained the rationale for your position, but she continues to argue the point.

#### Recommended Solution

*Do not* get into an argument in front of the class. As soon as you see that this individual intends to make an issue of the disagreement, you can say something like "I'm pretty sure that (whatever), but I could always be wrong. Why don't we check it out during the next break."

Even if you are 100 percent sure you are right, you have taken some of the edge off Nancy's argument by allowing that she may be right. And of course if she *is* right, you will come across the wiser for having allowed that possibility and not having persisted in arguing the point before you could research the answer.

Be sure to inform the class when the issue is resolved. If Nancy is right, give her credit and thank her for pointing out the error. If you are right, avoid flaunting that fact—remember, this is a win-win encounter. You can still thank Nancy for raising the issue and allowing you to verify the answer.

Remember that the audience will be more interested in WHAT is right than in WHO is right.

### Bad Attitude

#### Problem

Sam Sourface seems to have a bad attitude about being in class. He responds to your lecture with scowls and grumbles; he can be heard making comments to others like "This is a waste—I don't know why they sent me to this class."

*Recommended Solution*

Catch Sam during the very next break and—privately—explain to him the effect his negative attitude is having on the class. Take a few minutes to listen to his complaint, and be sincere in your effort to help him resolve it.

It may be that he is just bored with topics he already knows, but needs to stay in the class to pick up the the information or skills he doesn't know. Be understanding, and try to find ways for him to contribute in class what he knows that the others don't; point out how valuable that kind of contribution is. This way you have helped Sam feel he is part of the class and not alienated from it.

If Sam really does not need to be in class, suggest that he drop out and make it easier on everybody concerned. Treat him respectfully, but be firm: you will not tolerate his bad attitude during class.

## The Derailer

*Problem*

Debbie Derailer has a habit of getting you off track with irrelevant comments and questions that go beyond the scope of the course.

*Recommended Solution*

Hopefully, you explained the scope of the course during your introduction, so your audience should know what the class covers and what it doesn't. When Debbie tries to sidetrack you with her questions and comments, gently remind her "That's beyond the scope of what we're dealing with in this class, but if you'd like to discuss that point, I'll be glad to get together with you after class." Be careful about giving up your breaks to discuss things that can wait until after class that day.

Perhaps Debbie's questions are just poorly timed—she might be ahead of the class and asking about things you plan to cover later. In that case, mention that you'll be addressing that point later, and ask her to remind you at that time of her question.

## The Challenger

### Problem

Charlie Challenger continually challenges your credibility with questions and comments that seem to be designed to test your knowledge and experience rather than to enhance his learning experience.

### Recommended Solution

Like Sam, Charlie should be approached during the very next break and made aware that his behavior is having a disruptive effect on the class.

Remember that you know enough to be teaching the class, so even if Charlie really does know more than you do, it is not appropriate for him to be pointing it out in class. If he knows that much, he may not be gaining anything by being in the class, so suggest that he is free to withdraw.

If Charlie has some personal reason for trying to make you look bad, make a sincere effort to resolve the problem. Point out that neither he nor anyone else will be able to learn anything if he destroys your credibility. Try to keep him with you rather than against you.

Often, people who behave disruptively are in need of attention, and your private talk during a break might be just enough to get him off your case.

Remember, you are in charge, and you have the authority to do what is necessary to have a successful class. Maintain your confidence, and refuse to buy into the efforts of others to undermine your credibility.

## The Napper

### Problem

Sally Sleeper has trouble staying awake in class.

### Recommended solution

Don't take it personally when Sally falls asleep in your class. And don't embarrass her by making a joke or putting her on the spot with a question. Unless she is disturbing others, or most of the participants are also asleep, it is her problem more than yours.

If you notice people getting droopy-eyed, you can help them become alert without putting them on the spot.

You can raise your voice at an appropriate point in your lecture.
Move around the room so that the direction of your voice changes.
Mention the name of the person next to a sleeper, but in a non-threatening way; for example, "Tim, what was the problem you mentioned earlier that you had with your spreadsheet?"

Of course, if you often notice people falling asleep in your classes, think about what you can do to prevent that problem. Maybe you need to be more interactive so the participants are not just sitting and listening; maybe you need to try different teaching methods. See "Effectively Teaching Technical Material" for ideas.

Again, it is rare that your adult audience will really give you a lot of trouble. But you may be called upon to exercise your authority just to keep the class on track, and that is something you will want to become comfortable with. Confidence comes with practice, so hang in there and learn what works best for you in each type of situation.

## THE ART OF ASKING QUESTIONS

An excellent way to get your audience involved and make your class interactive is to ask questions. Just as there are different reasons for asking questions, there are also different ways of asking questions. The following list matches your purpose in asking a question with a description of the type of question you can use in that situation.

To obtain information, such as when you ask participants to introduce themselves, you ask a *factual question*. This kind of question can be directed to an individual or to the whole group and does not require a "correct" answer.

*"What types of software are you using with your PCs?"*

To obtain feedback, to see whether people are with you, whether they understand, whether they need a break, ask a *survey question*. This kind of question may be directed to an individual or to the whole group.

*"Do you have any questions?"*

*"Brian, you have a puzzled look on your face—do you have a question?"*

*"Is everybody with me?"*

*"Are you ready for a break?"*

Remember here to maintain good eye contact while you are waiting for an answer. When you ask if there are questions or if everyone understands, people need time to think before they answer. A rule of thumb here is to pause eight seconds after asking a question.

To get a variety of opinions or ideas, you can ask a *relay question*, which is a question directed to each individual. Again, you are not looking for "correct" answers but rather opinions or ideas. You can use this kind of question in a guided discussion, where, for instance, you have each person answer as you write the answers on a flip chart or the board.

*"What are some applications for which you could use a spreadsheet?"*

To involve an individual who you know can answer the question, you would ask a *direct question*. AVOID using a direct question to test knowledge; that can put someone on the spot unnecessarily and make him less likely to interact with you.

*"George, how did you finally get that command to work?"*

can get George involved by asking him to share his experience.

*"Glenda, explain how the KSDS index is structured"*

is testing Glenda's knowledge, and may put her on the spot if she is unsure or wasn't paying complete attention.

To stimulate thinking or test comprehension, you can ask an *overhead question* of the entire group. You may be looking for a specific answer or for ideas, and anyone may answer. People who are unsure or tuned out will not be put on the spot by this kind of question.

*"What kinds of things should you keep in mind when designing a screen?"*

*"Who can explain the structure of a KSDS index?"*

*"What is the difference between 2nd and 3rd normal forms?"*

To help a participant answer her own question, you can *reverse the question*—that is, direct the question back to the person who asked it. Reverse the question only if you are fairly certain that the individual will be able to think of the answer.

*"That's a good question. What do you THINK will happen here if you press the F2 key?"*

To encourage interaction from those who know, you can *redirect a question* to the entire class. This is an effective way to get help when you don't know the answer yourself, but it is also a good way to let those who know help those who don't, even when you do know the answer.

*"That's a good point. ARE there any times when you would WANT to use a GOTO?"*

*"Gee, I've never run into that myself. Does anyone know what to do when your password expires?"*

To assist those who don't know, ask a *leading question*, which offers two or more choices. You are asking which of the choices is correct, and anyone may answer.

*"At this point is the next step to obtain the next record or obtain the owner of the set?"*

To bring out different points of view, ask a *controversial question*. This is directed to the entire class, and anyone may answer.

*"Should the company stay with IDMS or switch to DB2?"*

Here are some things to keep in mind, with any questions you ask.

**1.** If you are leading a guided discussion, plan your questions, but be flexible enough to continue on the basis of the responses you receive.

**2.** Make sure the question can be answered if you are directing it to an individual.

**3.** Make your questions specific, relevant, and purposeful.

**4.** Use language the group understands.

**5.** Ask in a friendly, supportive tone. Remember, you want people to respond, not feel threatened.

**6.** Always pause long enough for people to respond to your question. Eight seconds is recommended—maybe longer when you are asking "Are there any questions?."

**7.** At all times, avoid putting anyone on the spot with a question.

## RESPONDING TO ANSWERS

Just as there is an art to asking questions, there is an art to responding to answers. The sequence does not end when your question has been answered, but when you have acknowledged that response.

The person answering the question wants feedback from you indicating whether the answer is what you are looking for. How YOU respond will affect not only the individual answering your question but the entire class.

Your goal is to encourage people to answer questions. If you respond negatively or do not respond at all, people will be less comfortable offering answers.

Here are some tips based on the type of situation you are responding to.

### The Answer Is Correct

If the answer is correct or—in the case of ideas or opinions—what you are looking for, acknowledge that fact. You will not only provide positive feedback to the person who answered, but you will also be letting the rest of the class know that the answer was correct. In addition, when others see your positive response to answers, they will be more likely to volunteer answers themselves.

Some positive responses on your part:

*"Right!"*

*"That's it."*

*"Good answer."*

### The Answer Is Close or Incomplete

If an answer is on the right track or partially correct, encourage the person answering to continue. Again, you are providing positive acknowledgment and also encouraging the participant—and the rest of the class—to think through an answer.

This kind of stretch reinforces learning, so resist the urge to provide the answer yourself or fill in the missing pieces if there is a chance that someone can get it. This is an example of where you might redirect the question to the rest of the class if the original answerer cannot get the complete answer. Here are some ways you can respond.

*"You're on the right track . . ."*

and then pause to allow the individual to continue. Maintain eye contact to emphasize that you want that person to try to finish answering.

If the answerer indicates that he has no more to add, acknowledge that fact; then make eye contact with the rest of the class to indicate that the question is open again to everyone.

*"That's right—go on . . ."*

to encourage a more complete answer. Again, maintain eye contact with the person you are coaxing.

*"That's a good start—does anyone else know?"*

to acknowledge and redirect the question.

### The Answer is Incorrect

Be especially careful how you acknowledge an incorrect answer or answers that aren't quite what you're looking for when you ask for opinions or ideas.

You want to let people know that you are pleased they tried to answer, and you want to avoid discouraging them from trying again. Usually you will redirect the question in this case, so be sure to acknowledge the original answerer before you open up the question to others. Here are some ways you might respond.

> *"Thanks, Ted, but that's not quite what I was looking for. Can anyone else answer?"*

Be sure you have acknowledged Ted with a friendly tone and good eye contact before redirecting the question.

> *"Good try, Mary, but not quite right. Anyone else?"*

Be genuine in saying "good try," and, again, be reassuring to Mary with your eyes and your tone when you acknowledge her answer.

### The Answer is Unclear

Sometimes an answer is unclear, so give the answerer the benefit of the doubt and ask for clarification before assuming the answer is incorrect.

> *"I'm not sure I follow you—can you rephrase that?"*

As always, use a respectful tone in questioning the answer you don't understand.

### Nobody Answers

If nobody answers your question, it could be that, truly, nobody knows the answer. It might also be that the question was not clearly understood.

Before assuming that nobody knows the answer,

Rephrase the question.
Offer a hint.
Guide the class through some questions you know they know the answers to, in an effort to lead them to the answer you are looking for here.

If there is still no response, let the class know that it is OK if they don't know—that is, don't berate them for not knowing. Maybe you need to review the material again.

If you have a good rapport with the class and know that humor will be taken well, you may kiddingly coax the class with your response:

*"Aw, come on—you know this, I know you do. Think about the problem Betty had in workshop yesterday. Remember? How did she solve her problem? Right! So how can you apply that solution to this situation?"*

## General Tips

Here are some general tips to follow in any situation where you are responding to somebody's answer to your questions.

**1.** Always pause long enough that people have time to respond.

**2.** Do not interrupt when someone is answering you; listen to the entire answer. If you continually cut people off, they will stop bothering to answer.

**3.** Acknowledge all responses, even if it is just with a nod or a look.

## RESPONDING TO QUESTIONS

In terms of providing feedback and positive reinforcement, responding to questions can follow the same guidelines as responding to answers. Of course, there are other considerations, too, when you are expected not just to respond, but also to provide an answer.

The following tips are based on the types of situations you may find yourself in.

### You Know The Answer

If you know the answer and this is not a situation where you want to redirect the question to the class or generate a discussion before answering, answer the question simply and directly.

The trick here is to answer exactly what's being asked. When you know a subject thoroughly, it is tempting at times to tell all you know. But telling all is not only inappropriate in the majority of cases, it can also overwhelm your audience. It's like answering your child's questions about sex: if you offer too much, you over-whelm, and understanding is lost.

If the answer is lengthy, or if you see puzzled looks, it is a good idea to follow your answer with "Does that answer your question?"

### You Need Time to Think

At times, you will need time to think before you give an answer. Even when you know the answer, you may need time to think about the best way to deliver it completely, clearly, and concisely.

It's OK to think. What you want to do, though, is let the audience know you are thinking—so they don't think you are just not answering—but avoid thinking out loud. This means

> *Look up with that universal "I'm thinking" look;*
> *OR*
> *Say that you're thinking—"Let me think about that," or "I'm thinking . . .";*
> *THEN*
> *Be quiet and think;*
> *BUT*
> *DON'T stand in front of the class and deliver a steady stream of "um, ah, well—let's see, uh," or verbally express random thoughts as you try to get your answer together.*
> *REMEMBER*
> *There is nothing wrong with a few seconds of silence.*

### You Want to Encourage Others to Answer

Sometimes you know the answer, but you also know that members of your audience may know the answer, too. By redirecting the question (see "The Art of Asking Questions") and allowing some-one else to answer, you are giving someone the chance to share her knowledge; you are making the class more interactive; you are adding variety (you're not always the only one answering ques-tions); and you are helping to keep the participants alert by ex-pecting them to be involved.

Other times, a question may be a perfect lead-in to a discussion you would like to generate. In getting the discussion going, you may answer the question first or you may defer answering; but either way, you open up the issue for discussion.

> *To the person asking the question: "I'm glad you brought that up, Charlie."*
>
> *To the class: "What are your thoughts about that?" or "Let's discuss that for a minute."*

Make sure that whatever you do, the original question does get answered.

### You Don't Know the Answer

Of course, there will be those times when you just don't know the answer, and that's OK. The best thing to do is to say that you don't know and then get the answer as soon as you can and get back to the class with it.

For a detailed discussion of this situation, refer to "When Your Answer is 'I Don't Know'."

### The Question Is Unclear

If you are not sure you understand the question or just what the questioner is looking for, do not hesitate to ask for clarification.

> *"I'm not sure I follow you. Would you mind rephrasing the question?."*

Or you might rephrase it yourself to verify that you understood correctly.

> *"What I think you're asking is . . ."*

It is far better to take the time to clarify than to spend time answering a question that was not asked. It's bad enough when that happens accidentally—when you don't even know that you've misunderstood the question.

### The Question Is Beyond Scope of Class

Questions that deal with subjects the course does not address can get you off track. And, they may be confusing or disrupting to others in the class who don't have the same knowledge or interest as the questioner.

Your goal here is to politely put off the person who asked the question without putting him down. If you have clearly stated the objectives of the course during the introduction, anything you say now will be like a reminder of the fact that there are some things you do not intend to cover, and the person asking the question is less likely to feel personally put off.

> *"That's a good question, Roy. But since it gets into an area we don't cover in this course, I would prefer to discuss it with you after class or during the break."*

### The Question Is Premature

Often, a question will relate to a topic you plan to cover later in the course, and to answer it now would be inappropriate or confusing. Again, be courteous in putting off the questioner.

> *"That's a good question, Theresa, and we'll be addressing that very issue after lunch. If you don't mind, I'd like to hold off answering that until we've covered utility commands. Would you make a note of your question and remind me then? Thanks."*

### General Tips

One really important tip that relates to all the above situations is this: always listen patiently as a question is being asked. Let the asker complete his question before you begin your answer, and remember to really listen to what is being asked—don't jump to conclusions when you hear the first three words. You'll end up answering what you expect to hear rather than what was really asked.

All the above examples are just suggested ways of dealing with the situation. As with everything else in teaching, you want to develop your own style and do what you are comfortable with.

## *WHEN YOUR ANSWER IS "I DON'T KNOW"*

If you don't know the answer to a question, there is nothing wrong with admitting that. Realistically, no instructor has ever been expected to know absolutely everything, and you won't be the first. But, how you handle not knowing can affect your credibility, so it's a good idea to practice how you plan to deal with such situations.

### *Why Don't You Know?*

One thing to consider when you get a question you can't answer is whether you could have been better prepared. If you are saying "I don't know" frequently, it may be that you do not know your subject well enough. In that case, make plans to increase your knowledge or experience level before you teach this subject the next time.

If the questions relate to things you probably could not have known, or would not necessarily be expected to know off the top of your head, chalk it up to experience. Remember to make a note of such questions for future reference—you may want to do some further research, or you may have occasion to use that answer again.

### *What Do You Say?*

When you have to admit that you don't know an answer, you have four basic courses of action:

**1.** Say you don't know, and promise to get back with an answer.

*"Gosh, I don't know. Let me check on that and get back to you."*

**2.** Say you don't know, and redirect the question to the audience, if that is appropriate, to see if someone else has the answer.

*"I don't believe I've ever encountered that problem before. Does anybody know how to get around that?"*

**3.** Say you don't know, and ask someone in the class to get the answer and report back to the class. This approach is appropriate when the question relates to something the class will be experiencing during an exercise or workshop.

*"Hmm, I've never seen it done that way before. Why don't you try during workshop and let us know."*

**4.** Say you don't know, and look up the answer on the spot, if that is appropriate; that is, if you need the answer right away and/or you have the necessary references on hand. If looking it up will be time-consuming, defer it until a break.

*"That would be in the user's guide. Let's look it up now."*

### What Do You Do?

**1.** Be confident. If you are confident in your not knowing (think about that), you can maintain your credibility despite the most appalling ignorance.

Notice in the examples above, that nowhere are you apologizing for not knowing or telling people how stupid you are, that you "should have known," and so on. If you believe it is OK to not know, your audience will believe it, too.

**2.** Get back to the class with any answers you promised. All the confidence in the world won't completely protect your credibility if you neglect to come back with the answer. This kind of professionalism receives high marks on participant evaluations.

### CONDUCTING A HANDS-ON WORKSHOP

Anyone who has tried to learn a skill knows that actually practicing what it is you are learning is the only way to master it. For that reason, technical classes that teach skills are much more effective when they provide practice through "hands-on" exercises or workshops.

In technical data processing classes, hands-on workshops are generally sessions in which the learners use the computer to practice the skills they are learning. The following guidelines are designed to help both you and the learners get the most out of your workshops.

**1.** The first step in conducting an effective hands-on workshop is to have a well-designed workshop to begin with.

Have clear learning objectives for the workshop.

Make sure the workshop is relevant to the objectives of the course.

Make sure the workshop uses only those skills the participants have already learned.

Avoid doing too much in one workshop; better to have smaller workshops that build on one another.

Practice: go through the entire workshop yourself to be sure the instructions are clear and accurate, that you've allowed enough time, and that there are no unexpected problems with the flow or the results.

**2.** Plan for your workshops to immediately follow the lecture that deals with what learners will be expected to do during those workshops.

For example, if you are teaching a JCL class, it is more effective to teach the JOB card and then have an exercise or workshop to practice coding a JOB card, than it is to lecture about JOB, EXEC, and DD cards before having the first workshop. By learning and then immediately practicing, learners are much less likely to forget key points and techniques before they get to use them.

**3.** Give clear instructions. Help participants know exactly what they are to do, using what resources and within what time frame, by walking through your instructions before the workshop begins. There is less chance for confusion or frustration if the instructions are written on the board or in a handout that participants may continue to refer to.

**4.** Be available to assist during the workshop. As people work through the assignment, they will discover what their questions are.

> Encourage participants to solve their own problems before you give them answers.
>
> Point out that it is OK to make mistakes—that the more mistakes you make, the more you learn.
>
> Pass along the problems/questions that one person/group has to the others in the workshop, where appropriate.

**5.** Pay attention during the workshop to find out

> How well your instructions were understood.
>
> What types of technical problems/questions arise.
>
> Whether the time you've allowed is appropriate.
>
> Whether your lecture adequately prepared participants for the workshop.

Use this feedback to refine future workshops.

**6.** End the workshop at the time you planned. It is important to keep the class on track, even when you are engaged in workshops. There will always be some people who could use more time, so if you are ending before everyone is finished, be sure to point out that there will be time to catch up during the next workshop, after class, or some other time.

Without that assurance, participants are likely to get frustrated and be worrying about finishing the assignment rather than listening to your next lecture.

**7.** After the workshop is ended, follow up and provide closure by discussing how the workshop went. Give participants a chance to share their experiences, and then summarize the points or techniques the workshop was designed to demonstrate.

A workshop that is well-designed and well-conducted is a tremendously valuable learning experience. It will often be the most outstanding feature of your class, so the time you invest in design and preparation will be time very well spent!

## CONDUCTING DEMONSTRATIONS

Demonstrating what you want your audience to learn is much more effective than simply talking about it or showing it in a picture. Especially when you are teaching others HOW TO DO something, you can get your message across more clearly, and probably in less time, with a demonstration.

As a teacher of computer-related courses, you will want to be comfortable demonstrating software and perhaps hardware, too. This section contains tips oriented toward demonstrating software, but the same principles can be applied to hardware demos as well.

Ideally, you will have some kind of large screen display of your terminal or PC screen so that the audience can easily see everything you are doing. But even without this special equipment, you can conduct an effective demo if you follow these simple guidelines.

**1.** Be prepared. Anyone who has ever conducted a demonstration on the fly will probably tell you that whatever can go wrong probably will when you have an audience. It's one thing if you can't answer a question, but if you can't even effectively demonstrate what you want others to do, your credibility can suffer. Some basic preparation will prevent this problem.

Define the objectives of your demo, just as you would for any other topic in your course.

Document the exact steps you plan to demonstrate. Then use this written list to guide you as you conduct the demonstration.

Hold a walkthrough. Test your demo with a sample audience and then get their feedback.

**2.** Be sure that everyone can see what you are doing. If you don't have a large screen display, be sure that whatever you want people to watch is positioned where it is easily visible.

**3.** Lighting plays an important role in visibility. Even if you have the large screen, you may need to dim the lights in order for the display to be easily read. Your goal is to make your display easily visible without making the room so dark that your audience cannot take notes (or stay awake).

**4.** Plan the seating so that everyone has a clear view of the demonstration.

**5.** Follow your plan. It is very tempting to deviate, but be careful. Not only do you run the risk of encountering a problem you hadn't anticipated, but if you are jumping around trying different things on the fly, it may be confusing to the audience.

**6.** Follow the KISS method and Keep it Simple (Sweetie). Remember that people are trying to follow not only what they are seeing but also the method or procedure you are demonstrating and the logic behind it. If you include everything that can possibly be included, people will be overwhelmed by the detail; they will come away less likely to remember the basics they were supposed to learn.

**7.** Remember to explain what you're doing as you go along. Even if people can see your screen very clearly, it will not be intuitively obvious to them how you accomplished each thing they see.

Odds are, they cannot see the keyboard, so if you don't mention that pressing the PF4 key is what takes you to the next screen, nobody will ever know.

Explain each operation you are going to perform; describe what will happen as a result; then perform it. A variation on describing the outcome is to ask the class what they would expect to happen if it is likely they would know. This is a good way for them to review and integrate information previously presented.

**8.** Go slowly. Again, your audience is trying to comprehend not not only what they see but what you say and the logic behind the process.

Be especially careful as you move around the screen. It helps if you can make the cursor a full-character rectangle, as opposed to an underscore, and if you can make it blink. In any case, move the cursor slowly enough that people can follow it.

**9.** Pause after each step to let what you've done sink in.

**10.** Allow for questions as you go. When people get confused, they are distracted from hearing what comes next. Giving a demonstration should be like any other presentation method in that you want to make it as interactive as possible and make sure that people have grasped one concept before you go on to the next.

**11.** Consider letting the class follow along on their computers or terminals. If the demo is not just informational but is teaching your participants things you want them to be able to do, it may be worthwhile to let them use their keyboards and do as you do as you go along.

While this approach sounds very effective, it does have one major pitfall: if someone makes a mistake or gets hung up, the flow of your demonstration will be interrupted while you trouble-shoot the problem. Such interruptions will disturb the flow for the rest of the class as well, and you may be undermining the advantage of doing a demonstration in the first place.

The recommendation here is that you determine first the objectives of the demonstration. If you are using it in your presentation as a way of introducing new information, it is preferable for you to demonstrate and your audience to watch.

However, if you are using the demonstration as part of a hands-on workshop and want to step people through what they will be doing, it is appropriate for them to follow along on their machines. The understanding here is that you will expect to stop and help people as you go.

**12.** Summarize, as you would any other presentation, recapping for your audience the purpose of the demonstrate and the points you expected they would learn from it.

## General Tips

Demonstrations can be extremely effective, but if they are are not well planned and executed, they may provide little benefit and—at worst—can be a disaster.

Keep in mind, too, that a live demonstration need not be

with a computer. You might decide to use a demonstration as an analogy to illustrate a concept.

For example, to drive home the point that system specifications must be specific, detailed, and complete, you might ask the class to give you exact instructions for, say, preparing breakfast. You are prepared with the necessary props, and you do exactly as you are told. When your response to "Open the cereal" is to rip open the box and let the contents fly everywhere, you are "demonstrating" a point.

This kind of demonstration, like any other, will be most successful if it is

> Well-planned
> Simple
> Easy to see
> Easy to follow

## GETTING PHYSICAL

One way you can enhance a stand-up presentation is by getting physical—that is, moving your body. Watching an instructor who doesn't move can become tedious no matter how interesting the subject is, but an instructor who is animated and moves around brings life and energy to his presentation.

Once you have mastered the basic skills of communicating authority, control, and confidence with your body language (see "The Visual Message") you are ready to use your body to communicate more.

Here are some situations where body movement can add to what you are communicating.

### Keeping Things Moving

Avoid standing in one position for too long. If your participants must focus their gaze for long periods of time, not moving their eyes much less their heads, they could become droopy and inattentive.

Your goal here is to move around, but don't be so constantly on the move that it distracts your audience or makes them nervous to watch you. After standing in one spot a few moments, try taking a few steps to one side. Now stay in that new spot for at least 15 to 20 seconds before moving again.

Remember, too, that as you move, you want to avoid turning your back on your audience.

### Emphasizing a Point

Larger-than-life gestures can help draw attention to a point you are making. Be careful not to overuse this technique, however, for if you use too many grand gestures, they become routine rather than emphatic.

Another movement that can be used with the gestures or independently is to move toward your audience as you stress an important point. Maintain eye contact and a nonthreatening manner as you walk or lean forward. Again, avoid overuse.

### Making Your Audience More Alert

If you notice some glazed eyes and it's not a good time for a break, you might try walking around the classroom as you continue to talk. If you have room, walk around the side toward the back, so that heads must turn to follow you. If all you have is a center aisle, use that, and then stand in the back for a moment or two.

Between moving to watch you, hearing your voice from a different direction, and, for some, feeling your presence right beside them, participants are guaranteed to perk up if they have been droopy.

Naturally, you want to take your stroll at a time when it is convenient to leave the front of the room.

### Getting Comfortable

One way to communicate that you want everyone to relax and be comfortable is to remove your suit jacket. Typically, you will do this near the beginning of the class but after the introductions, during which you may want to maintain an air of formality by leaving the jacket on.

There is no need to make a big deal of this; you are in charge, so unless you have an unusually formal audience, you shouldn't have to ask permission to remove your jacket.

Note: *This tip applies to teaching a class. If you are giving a formal presentation, it is probably appropriate to leave the jacket on for the duration of your talk.*

Another way to communicate "Let's get casual," is to sit or partially sit (keep one foot on the floor) on the edge of the instructor's table, assuming you have one. Women, especially, need to be careful about maintaining a professional image if they sit casually. Sorry—that's just the way it is.

### Getting Down to Business

There may be a point during the class—at the beginning of a workshop or group discussion, for example—when you want to communicate to participants that "We are going to really get down to business now"; in other words, "Get ready to work."

Rolling up your sleeves at this point will convey such a message. If you don't have sleeves to roll up, you might try pushing the overhead projector aside and leaning authoritatively against the instructor's table. If you don't have these props to help you, you can always make the motion of rolling up your sleeves.

### General Tips

As with everything else you do in class, find what works best for you. Use motion as much as you can, but let your movements be natural and comfortable so that they truly enhance your presentation and don't simply distract your audience.

### TAKING CARE OF YOURSELF

We all know we can perform better if we are mentally alert, emotionally at peace, and physically rested and feeling good. But stand-up teaching challenges all those areas of well-being: it is not only mentally strenuous, it is also physically and sometimes emotionally taxing.

Naturally, you will be taking steps before your class to be ready for the mental, physical, and emotional demands, but there are things you can do during class as well. Remember, if you don't take care of yourself, nobody will.

## *Your Physical Body*

**1.** Dress comfortably. Avoid tight neckties and belts. Wear clothes that allow you to

Move
Breathe
Adjust to temperature changes

**2.** Take care of your feet.

Wear comfortable shoes! Rubber soles and padded insoles are recommended.
If your feet and legs get sore from standing all day, wear support socks or hose.

**3.** Go easy on your back. Allow yourself to sit occasionally. Have an instructor's chair up front for breaks and workshops. If you must move heavy things, ask some of your participants to help you.

**4.** Eat well. Don't skip meals just because you need to prepare for the afternoon session; eat and prep at the same time if you must, but don't leave out meals all together.

## *Your Mental Body*

**1.** Be well-prepared. Relieve your mind of a good portion of your worry by being well-prepared for your class. (See *"Preparing to Teach"*) But,

**2.** Don't be overwhelmed by preparation. If your class is more than a half-day, don't feel that you need to prepare for the entire class before it starts. Prepare each evening for the next day, and glance at notes during lunch and breaks if you must.

**3.** Make instructor's notes. Don't rely on memory, intuition, and luck to get you through *everything*. Having a good set of notes to rely on when you need them can help you mentally relax.

Along with your notes, it's a good idea to have the names and phone numbers of people you may need to call for help with questions or for technical support.

**4.** Be Positive. Assume that things will go as you want them to go. Remember, your thoughts can create a self-fulfilling prophecy, so why not create a positive outcome?

**5.** Take a break. Remember that instructors deserve a break, too.

Try to allow time to relax while the rest of the class is at break; this means you want to avoid filling your break time with class-related things that can wait until you are through for the day.

If you really need to wind down, be firm in letting participants who want to chat know that you need a few minutes to prepare the next lecture, to rest your voice, or to just be off-line. If you have questions to research or out-of-scope questions you said you would answer outside class, leave them for the end of the day if they do not require immediate attention.

### Your Emotional Body

**1.** Resolve any negative feelings before your class. If you have fears about teaching, positive visualization may help you put them aside. (See "Visualizing Success.")

**2.** Don't take it personally when things go wrong. Things just happen: people may arrive late; the power may go off; you may get a question you can't answer; you may have to rush to finish on time. That's the way it is—try to take such things in stride.

Even if a participant challenges your credibility, you don't have to buy into the negative implications. Assume that the the challenger really just wants to know whatever is being asked. Again, "see" yourself handling these situations the way you would like to see them go.

**3.** Take a few deep breaths anytime you feel yourself starting to feel frustrated or annoyed. No need to let things get to you. Teaching is fun, remember?

## EXPANDING YOUR KNOWLEDGE BASE

The best instructors seem to be a wealth of realistic examples and relevant stories. Certainly, some of this repertoire has come from the base of knowledge and experience these instructors had before they started teaching, but in most cases, the teaching experience itself has helped build that base.

Here are some ways you can use your teaching to expand your own knowledge base. Mostly, it is a matter of remembering and documenting what you experience.

**1.** Encourage participants to share their experiences. Adults like to share their experiences, and this is a good way to get your audience interactively involved. Plus, it gives you insight that can help you tailor your presentation as you go. AND as you listen, you might find some excellent—real-life—examples to share with your next class.

**2.** Document real problems that your participants bring up in class. Remember, adults tend to be problem-oriented, and if they have had problems back on the job that are related to what you are teaching, you can be sure that they will want to talk about them in class.

If someone in your next class asks you "How do you deal with . . . ," you may have an authoritative answer, even if you have never run into the problem yourself.

**3.** Save workshop exercises that may make good examples in another class. If a participant has come up with a unique solution, or even a never-before-seen colossal mistake, you may find it an effective teaching aid later.

### YOUR PROFESSIONAL PRESENCE

As an instructor, you are in a position of influence and authority, and with that position comes responsibility—a responsibility you may not be faced with in the nonteaching part of your job.

Here are some recommendations for maintaining your professional image as an instructor.

**1.** Be prepared. This will let your audience know that you take your job seriously.

**2.** Dress appropriately. Generally, this means dressing up, but the culture of the company where you're teaching will dictate what is appropriate. When in doubt, dress up, and if more casual dress is acceptable, you can always dress down after the first day.

**3.** Make only positive references to your company and to your colleagues—no matter how you personally feel! Stay neutral if others make negative comments.

**4.** Show respect for all participants. This means, among other things, that you want to be careful in your use of humor so that you don't embarrass anyone. *Always avoid* sexist, racist, and ethnic jokes.

**5.** Leave your personal problems outside the class.

### ENDING THE CLASS

The ending of your class is not nearly so critical as the beginning, but it is appropriate to provide some kind of closure to the learning experience. If you get to the end of a hands-on workshop and then start packing up as if the whole class is over, your audience may be confused about when it all ended.

Most of us appreciate a final chance to ask questions and review major points before we go out into the world with our newly acquired knowledge. The following list suggests some things you may want to include as you end your class; you be the judge of what is appropriate in your situation.

**1.** Review major points. Just as you have summarized at the end of each major topic as you taught the class, you also want to review at the end of the class. A summary/review is a good way to

Find out what people still don't understand.
Remind the class what points you feel are important for them to
    remember.

**2.** Review objectives. One way to review is to go through your list of objectives again and point out what you have covered during the class that meets each objective. This is a way to summarize, and it also reminds your participants that you have, in fact, met the objectives you set out to meet. In the event that you have missed meeting any objectives, this is a good way to find that out before the class is over.

**3.** Ask for questions. Give everyone a final chance to ask questions about the material you've covered.

**4.** Let participants know what comes next. If the class you just taught is a prerequisite for another course, you can remind participants that they are now ready for that next course. Or just let them know in general where they can go from here.

For example: "Now that you have completed Intro to DOS, you are ready for Advanced DOS. Or, you can take LOTUS 1-2-3, DW-4, or RBASE 5."

**5.** Mention how much you have enjoyed the class, assuming, of course, that such a comment is sincere.

**6.** Remind everyone to fill out an evaluation. Allow about 10 minutes for evaluations to be filled out before people leave. You are more likely to get more evaluations back if they are filled out in the classroom rather than mailed back to you later.

You may prefer to leave the room while people are writing their evaluations.

**7.** Let people know how they can reach you or other appropriate people with questions following the class.

**8.** Collect the evaluations. Since most evaluation forms give the evaluator the option of identifying her/himself, you can help protect anonymity by not reading any evaluations while there are still participants in the room.

**9.** Leave the room the way you would like to find it the next time you teach. Leave things clean and orderly, and make sure all electrical equipment is turned off, unplugged, packed up—whatever is appropriate.

**10.** Take all your things with you.

If you have taught before, you know that there is really more to ending a class than just ending the presentation with some or all of these 10 points. At some appropriate time, you want to do the following as well.

**1.** Read the evaluations. You needn't read them the minute everybody leaves—you might want to wait until the next morning or until after the weekend if you are wrapping up on a Friday. If you give yourself a little distance from the actual class, you may be in a better frame of mind to read the criticism along with the praise.

**2.** Evaluate suggestions for improvement that are documented in the participant evaluations.

Sometimes there will be good suggestions that you have good reasons for not adopting—for example, "This class should be held off-site," when your training budget prohibits such a choice. If you know who made the suggestion, it is a thoughtful gesture to contact the individual and explain that, yes, it was a good idea, and then explain why you are unable to act on it.

At other times, the suggestions will be exactly what you need to improve the course.

**3.** Refine materials, notes, visual aids. Especially if the course is new, you will find things that you want to change before the course is taught again. Plus, you'll have input from the participants' evaluations. The best time to make these changes is usually right after the class is over, while your thoughts are fresh.

Typical changes that you want to think about include:

Improving clarity on visual aids, handouts, workshop instructions.
Updating content to keep pace with changes in the technology, environment, or expected application of the material.
Correcting errors in typing or content.
Adding depth based on the new things you've learned from having taught the course.

**4.** Act on personal improvement suggestions. The evaluations, as well as your own experience, will give you clues as to the personal changes/refinements you can make in order to have a more successful class next time.

Do you need or want to change/improve any of the following?

Your knowledge of the subject—do you need to learn more or gain additional experience?

Your knowledge of the course—would more familiarity or practice make the next class go smoother?

Your preparation—could you spend more time preparing yourself, the classroom, your materials? Could you be better organized?

Your presentation skills—is additional training or practice indicated?

Hopefully, when it is all over, you will feel you have had a good time, you will have learned some things you didn't know before, and you will be enthusiastic about teaching again!

*Appendices*

# *Walkthrough Evaluation*

*Instructor:* _____

*Course:* _____

For each item below, use one of these symbols to indicate your response:

___ + ___ You agree, or the activity was handled effectively.
___ – ___ You disagree, or the activity was not performed.
___ * ___ You recommend improvement (please fill in comments).

### *INTRODUCTION—The instructor:*

1. ___ Introduced him/herself.
2. ___ Had participants introduce themselves.
3. ___ Clearly stated course objectives.
4. ___ Gave overview of course.

Comments: _____

_____

_____

### *PRESENTATION—The instructor:*

5. ___ Demonstrated good knowledge of subject matter.
6. ___ Was well-organized.

7. ____ Was well-prepared.

8. ____ Managed time well.

9. ____ Controlled class flow (discussions, interruptions, transitions between topics).

10. ____ Encouraged questions.

11. ____ Answered questions confidently and effectively.

12. ____ Encouraged participant interaction.

13. ____ Kept participants' attention.

14. ____ Maintained good eye contact with audience.

15. ____ Communicated effectively (volume, speed, articulation, language appropriate to audience)

Comments: _____

_____

_____

### TEACHING METHODS—*The instructor:*

16. ____ Used visual aids.
    a. ____ Aids were clear, easy to understand.
    b. ____ Aids enhanced the presentation.
    c. ____ Aids were used appropriately and effectively.

17. ____ Used exercises or hands-on workshops.
    a. ____ They supported the objectives of the course.
    b. ____ Instructions were clear.
    c. ____ Instructor offered appropriate assistance.

18. ____ Used interactive discussions.
    a. ____ Discussion was relevant.
    b. ____ Instructor kept on track.

19. ____ Used on-line demonstration.
    a. ____ Demo was relevant.
    b. ____ Demo was well-organized and easy to follow.

Comments: _____

_____

_____

20. Overall impression of the walkthrough:
    ____ Excellent
    ____ Very good

_____ Good
_____ Fair
_____ Poor

**21.** Suggestions: _____

_____

_____

_____

# *Tips for Vendor Instructors*

*I*f you teach for a company that sells education and training classes, you will face some issues that your in-house counterparts may never encounter.

Here are a few tips that may help you be a successful vendor instructor.

**1.** If your company has made the arrangements for you to teach at a client site, it is a good idea for you to call the client yourself and find out some things about the client and about your audience. This also gives you a chance to begin establishing a good rapport with the client before the day of the class.

Here are some general things you will want to ask:

    **a.** Exact *directions* to the building *and* to the classroom where you will be teaching.

    **b.** Details about *accessing the room*—will you be able to get in early? Will you need a key?

    **c.** *Security* (for instance, at official government and military sites)—will you have to make arrangements to be met at the door and escorted?

    **d.** Client's expectations with regard to *class hours*, lunch, and breaks.

    **e.** *How many participants* will be in the class.

    **f.** The *makeup of the audience*—systems personnel, end users, managers, etc. What is their experience level with what you are teaching? Are there any politics it would help you to know about?

g. Client's expectations with regard to *scope and depth of the course content.*

h. Any *tailoring needed?*

i. If the *participant materials* are being mailed, have they arrived?

If your class includes hands-on workshops, here are some additional things you will want to know.

j. *What kind of mainframe computer or PCs* will you be working with?

k. What *operating environment* will you be interfacing with? (operating system, text editor, etc.)

l. What *software release(s)* will you be working with?

m. If you have special setup needs (offloading tapes, establishing user ids, etc.), *who will be available to help you?* Get a name and a number.

n. *Who do you call if there are technical problems* during class?

o. *Where are the terminals/PCs* that will be used during workshops? (in the classroom, on another floor, etc.)

2. Once you arrive at the client site, you will want to check these things.

a. *Who is your contact* in case you have questions about the equipment, need a photocopier, etc. How do you reach this person during class?

b. *Did the participant materials arrive?* Are there enough copies?

c. *Room arrangement and equipment*—see the "Preparation Checklist" in Part 2.

3. Remember, as you interact with your clients both in class and outside class you represent not only yourself but your company.

a. Your clients will expect you to be *knowledgeable about your company*—to know its products and services and something about its previous history with this client.

b. Avoid saying or doing anything that will *compromise your company's effectiveness* with this client.

c. Refrain from saying anything derogatory about *your company's competition.*

d. Unless it is your job to settle such things, avoid getting involved in discussions in which the client is complaining about *past experiences* with your company.

    **e.** Carry with you a few copies of *your company's literature*: course catalogs, class schedules, product fliers.

    **f.** Be generous in handing out *your business card.*

**4.** Know *who in your own company you can call* if you have questions you cannot answer.

**5.** Let class participants know whether it is appropriate for them to *contact you after the class,* or if there is someone else at the company they can/should call with questions. Be careful not to offer for free help that your company normally charges for.

**6.** If you have time, *get to know the training staff* (if there is one) at the client site. This is a good opportunity to network.

**7.** Following the class, you may find it appropriate to *stay in touch* yourself with the client, especially if you expect to teach there again. This will allow you to get more feedback about the effectiveness of your training and will help you maintain a good rapport.

If you happen to be a traveling instructor, as many vendors are, here are a few additional tips.

**1.** Assuming you are flying, plan to *arrive enough ahead of time* the night before the class to allow yourself a good night's rest.

**2.** If you arrive early enough, you may want to *take a test drive* (or test walk) to make sure you can find the client site.

**3.** Invest in a *sturdy travel bag* that can comfortably accomodate all your materials—instructor guide, technical manuals, boxes of transparencies, markers, company literature, etc.

**4.** It is also worthwhile to invest in a *fold-up luggage carrier* to carry the above-mentioned bag through airports with the rest of your luggage.

**5.** *Travel light.*

# *Tips for Volunteer Instructors*

*I*f you are a volunteer instructor—that is, a full-time technician who teaches occasionally—your experience will be somewhat different from that of a full-time instructor.

Here are a few tips that may help you be a succesful volunteer instructor.

**1.** Don't be shy about *asking for help* with your course development and presentation skills. If you are assisting a full-time training staff, they will probably be more than happy to train you as an instructor.

**2.** Be open to *feedback*—it will help you improve.

**3.** Give yourself *realistic expectations*—strive for excellence, not per-fection. If you teach only occasionally, accept the fact that it will be harder to refine your skills than if you were a full-time instructor.

**4.** *Observe other instructors* when you can.

# *Networking*

*A*s a professional instructor, you can benefit by networking with other professionals. Whether you are an in-house instructor or a vendor, networking can help you

Share ideas/test your ideas.
Find solutions.
Stay current.
Have a shoulder to lean on when the going gets rough.

If you are new to professional networking, the following tips may help you get started.

**1.** Subscribe to *professional periodicals*—know what is happening in your field; get to know the subject authorities and trend-setters.
*Data Training Magazine* is one publication that caters specifically to the needs of data processing trainers.

**2.** Join a local *consortium*: most large metropolitan areas have a professional technical trainers' organization. Typical benefits include:

Joint sponsorship of vendor classes, which individual members may not be able to justify hosting on their own.
A forum to resolve common training issues.
Informative programs that help you stay current.
Networking!

*Data Training Magazine* regularly publishes a list of training organizations throughout the country.

**3.** *Keep track of names and phone numbers* of other professionals whom you would like to stay in contact with. Even if you talk to each other only once a year, it is worth keeping in touch with others who do what you do.

**4.** Be prepared to *give as well as receive.*

# *Instructor's Diary*

### Monday, April 5

Found out today that Al is transferring out of the group to become a systems programmer—his last day is the 16th.

Barb [our manager] said she'd like me to teach the Logical Database Design class Al was going to teach at our Washington, D.C. data center next month (May 4–7). Told her I would look at what was involved and let her know by tomorrow. (I've never taught database design before! I'm nervous about doing it, but on the other hand it will be a good professional stretch. . . .)

### Tuesday, April 6

Told Barb that I would like to do the class, but it will take more prep time than it would to teach a class I've done before. She asked how much time—I said I would meet with her tomorrow to go over my time estimate.

I hate having to put those things in writing. But since I am tied up with the entry-level programming trainees for another 3 weeks yet, I know I need to think about this carefully and make sure I really do have time to fit it in.

### Wednesday, April 7

Met with Barb. Gave her the following estimate of my time on the design class. Hope this is accurate, because I'm so busy with the trainees that I won't have much more time than what I've estimated. This is what I was able to throw together last night after the kids went to bed:

*½ hour*

to call the manager in D.C. [Vic] who requested the class and find out what the audience is expected to be.

by 4/9

*2 days*

to review the course. I've sat in on Al's class before (3 months ago), and I know logical design, but I'm just not familiar with how the course flows. Need to go thru materials, instructor notes, etc.

Probably 4 ½-day sessions, between 4/8 and 4/16.

I'll fit this in during the afternoons, when the trainees are working with their mentors on their final projects.

*2 hrs*

to meet with Al before his transfer.

4/12, 1 PM

(I checked with him this morning)

*1 day*

to practice with the data dictionary, since the class includes teaching how to enter the data elements into the dictionary. I know the dictionary, but I haven't signed onto it since Al's last class, and I know there's a new release since then.

4/20, afternoon.

Gail, in tech support, is my contact if I need any help.

*3 hours*

to read all the prework reading we assign and to review references in the technical manuals.

<u>4/21, evening.</u>
Dan [my husband] is taking the kids to the circus that night—good time to concentrate.

*½ day*

to practice the exercises and the data dictionary workshop. I need to have all the details of the case study fresh in my mind, because it is so complex. (Luckily this is the same case study we use in some other database classes, so I am familiar with all the background stuff—just not the details for the design workshops.)

<u>4/23, afternoon.</u> Graduation is that morning, so I should be able to have the afternoon free. . . .

*2 hrs*

for a walkthrough. Have already scheduled a classroom for the *26th from 9:00 to 11:00.* 3 other instructors said they could sit in, and Al may be able to drop in, too.

*½ day*

to change/add/delete anything as a result of the walkthrough.

<u>4/28–4/29—</u>
I'm at an off-site meeting the 27th.

*1 day*

to wrap up the course and make any revisions after teaching it once.

<u>5/10, all day I hope.</u>

### Thursday, April 8

Had hoped to spend the afternoon beginning to review the design course, but got called into an unscheduled staff meeting that took most of the afternoon. That left me with about an hour before leaving, and I was barely motivated at that point in the day.

Took the instructor's guide home; fell asleep reading the first section. Oh, well.

### Friday, April 9

FINALLY got thru to Vic in Washington, after 3 days of telephone tag!

He says the class will probably be 14 people: 8 systems people (programmers and analysts) and 6 business clients, all from the same high-priority project team. Vic is viewing this class as a pilot for training all team members at the same time in data analysis and design. (AGHH!) Of course, I told him how much I was looking forward to the opportunity.

Fortunately, I composed myself in time to ask some details about the project (thank you, God), and I did remember to take notes. We talked a little about my experience with database design, and it occurred to me that the last project I implemented before becoming an instructor was very similar to his project, in terms of the file they want to convert and the functions they want to support.

Thank GOODNESS he didn't ask me how often I had taught this course! I caught myself a couple of times starting to say something about that, but luckily I realized in time that he didn't need to know that. He has confidence in my ability with the subject, and that makes me feel good. In fact, he says he's glad I'm the one teaching it BECAUSE of that experience, because he knows I'll teach his team the best way to do things.

That's nice, but now I have some high expectations to live up to. Plus he considers this a pilot, so I know he'll be paying close attention to feedback. (Nothing like a little pressure to keep me on my toes!)

The good news is that this morning's class with the trainees ended earlier than I expected, so I had some time to get into the instructor's notes—finally! I can see I've got my work cut out for me.

### Monday, April 12

Had a challenging class this morning with the trainees, then rushed thru lunch in order to meet Al at 1:00, only to find a note that he had an "automotive emergency on the way back from lunch," as he put it, and we would have to reschedule (story of my life!)

Spent the afternoon reviewing the course instead, and came up with some questions I wouldn't have known about if we'd met earlier.

We finally ended up meeting at 4:30; I was able to get hold of Dan at work—not an easy task!—and ask him to pick up the kids. He wasn't thrilled about it, but I suggested that this would give him and the kids an opportunity to make dinner together and surprise me, since I would be home late. I WAS surprised— they couldn't decide what to make, so they took me out to dinner!

Anyway, Al and I had a good meeting. He gave me some tips on timing and pointed out where participants usually run into problems in the workshops. Also, he said that he had taught a class before in which everyone was from the same project team, and that it was a really good experience for the team. BUT, he said, you have to be aware of the politics and be prepared to deal with it. Think I'll call the project manager tomorrow and find out what I can.

### Wednesday, April 14

My horoscope this morning said "Remain open, and all you need will be there for you." I think I'm a believer because I got to spend most of the afternoon reviewing the course, just as planned, plus Al was available for questions. I'm getting a good feel for the flow, the timing, and how the workshops tie together. I feel good about the class, and know I can handle it.

### Friday, April 16

Al's last day in the group; we took him out to lunch. He said I could feel free to call him if I have any more questions on the design course. I still feel kind of nervous knowing he's not right in the next office anymore.

Spent the afternoon finishing up with the instructor's notes. My actual time to get familiar with the course turned out to be about 3 hours less than my original 14-hour estimate. But next week is the heavy stuff—reacquainting myself with the dictionary, and making sure I understand all the data modeling and dictionary exercises.

### Friday, April 23

Well, the trainees have graduated, and I somehow managed to play around with the dictionary and the workshops this week. As usual, there were interruptions, and what was supposed to be 2 afternoons turned into pieces of 4!

Also got around to calling Art, the project leader on the systems side of the group that will be in the Design class. He filled me in on some of the concerns of his group and pointed out a pesonality conflict that exists between one of his senior analysts and the project leader from the client area. Sure glad I found out now. Haven't been able to reach Sara, the client project leader, yet.

Still haven't looked at the prework reading assignments—Josie [my 3-year-old] got sick Tuesday night, so the circus trip was cancelled, plus I was up two nights with her. I'm tired. I guess if worse comes to worse, I can do the reading on the plane. . . . I'm really feeling tired.

Haven't prepped for my walkthrough yet; guess I'll do that over the weekend. After I've had some sleep! Hope I haven't bitten off more than I can chew.

### Monday, April 26

Had my walkthrough today for the design course. Larry, Lucy, Katy, and Henry attended—Al couldn't make it. I went through my introduction and then practiced delivering what for me was the most complex lecture; then I had the audience go through one of the workshops.

It went pretty well, but I realize I still have a confidence problem with some of the lecture. It's one thing to design data-bases, but something else again to EXPLAIN HOW IT'S DONE!

Also, I got feedback that my workshop instructions were confusing, so I'm going to rewrite those. It's more work, but it really is better to learn these things now than during a real class!

Barb told me today that we've been asked to teach another Logical Database Design class the week of May 10th, and asked if I'd like to teach it. I enthusiastically agreed, pointing out that teaching two classes so close together would help me really feel comfortable with the course. After she left, I admitted to Larry that what I really wanted to tell her was that she was welcome to teach it herself.

### Thursday, April 29

Only 5 more days til I teach Logical Database Design! I've made the revisions I needed (took twice as long as I anticipated since we were short on clerical help this week—never fails!); the class materials have been shipped to the Washington office; I've got my plane tickets, my hotel reservation, and my rental car reservation; and I've told my friend Cathy that I'll be in town for a few days—maybe we can get together for dinner.

Sara finally returned my call today. She seemed very friendly and gave me some more background on the project. She didn't mention the personality conflict with Bill, the senior analyst, but I can tell from talking to her, and from knowing Bill, where at least part of the problem is—very different interaction styles.

LUCKILY, Dan was looking at my tickets the other night and discovered that they had me flying round trip to DALLAS, Texas, instead of DULLES airport outside Washington. I guess the guy in Travel misunderstood me when I asked to fly to Dulles! Thank goodness there was time to get new tickets.

I THINK everything is all set now. But I'm still nervous. Still haven't looked at the prework reading.

### Monday, May 3

I can't believe the day I've had! First, I was almost late getting to the airport—Dan was late picking up the kids and meeting me at work; then the flight left late; and then after we were finally in the air, about 20 minutes, an engine caught on fire, and we had to turn back.

The good news was that they put us all on the next flight, and because I was one of the last ones in line and the plane was full, I got to go first class!

The (next) bad news was that my luggage didn't arrive with me. Fortunately, I had my "teaching bag" on board with me, so I arrived with my instructor notes, transparencies, and manuals intact. Also fortunately, I never got time to change clothes before going to the airport, so my clothes will get me by in class tomorrow until my luggage catches up. (Clothes will be the least of my worries!)

Still, I didn't get to my hotel until around 10 PM, only to find that they had lost my reservation. I was so exhausted! Luckily, there were rooms in another hotel nearby, and since I had a car, I could easily get to it. Checked in; called Dan to let him know I was alive; found out Jody [our 5-year-old] had caught his finger in the car door when they got home and cried himself to sleep asking for Mommy. Dear God, should I really be here?

Finally, by 11:00, I was soaking in a hot bathtub, trying to think positively about tomorrow. Suddenly, this class seemed more like a huge mistake than like a professional stretch.

I kept thinking of that one lecture I knew I was nervous about, so I closed my eyes—hoping I wouldn't drown—and pictured myself going through that part of the course, confidently explaining everything in superb fashion and answering with ease any questions that came up. It was hard to imagine just what questions I might get, but I saw the end result of everyone thinking I was an expert of impeccable credibility. After about 20 minutes of visualizing a great class, it all seemed quite achievable!

Lying in bed, my mind came back to another reality—that of absentee parenthood. . . . I was too tired to cry for long.

### Tuesday, May 4

I ate breakfast in my room as I quickly skimmed the prework reading—forgot all about it last night. Didn't realize 'til I got to the Washington office that I had spilled orange juice on my skirt. I was more concerned about the hole in the toe of my stocking because I was sure with my big toe sticking out tightly thru the hole, circulation would be cut off and gangrene would set in before the first break.

Vic met me and showed me to the classroom—it had been about a year since I had taught a class there, but I felt at home. Sleepy, but at home.

It figures that one box of class materials hadn't arrived, so we had workbooks for only 12 of the 14 people. They were able to share until the other books arrived around 10:30. (*Sheesh!*)

I couldn't believe it! The first day went great! I guess everything that could have gone wrong already had gone wrong. I was on a roll and had a ball! Even managed to forget about the dirty skirt and my big toe, which really was painful until numbness set in.

My luggage was at the airport when I got out of class, so I had it delivered to the hotel by dinner time. Naturally, I had already spent a small fortune for deodorant and makeup at the hotel shop this morning, but as long as I have my clothes, I'm happy.

Talked to Dan and the kids. Jodie feels better, Josie has a pet spider named Clyde, and Dan was asked to go out of town tomorrow! He was able to postpone the trip 'til next week—thank goodness!

I told him all the things that had gone wrong, and he said something that really made me think: he said "I guess as long as the plane didn't crash, everything else is really no big deal." I thought about how all the really important things had gone OK and decided to quit feeling sorry for myself.

Stayed up late prepping for tomorrow's class. Visualized that lecture again—it comes up tomorrow afternoon—and fell asleep feeling much more peaceful this time.

### Wednesday, May 5

That lecture went great! Just as I visualized. And my background notes on Vic's project really came in handy; I was able to make comments and come up with examples that everyone—programmers, analysts, and client analysts, alike—could understand and relate to.

Knowing the situation with Bill and Sara, I avoided putting

them on the same workshop teams today, and I was able to use examples that made them both look good. I'm hoping they'll each come out of this with at least a healthy respect for the other's point of view.

I'm exhausted, but I'm pleased with how it's going. I'm up late prepping again tonight, after having taken a bit of a break to meet Cathy for dinner. I feel confident with the material, but I need to get on top of all the details of the case study exercises.

### Friday, May 7

What a day! The class was a huge success, and the participants enjoyed it so much, they took me out to lunch today. That really made me feel good.

But the rest of the day has been incredible! When I was returning my rental car, they had me pull into their gas station so they could fill the tank before I turned in the car. Just as I rolled down my window to hear what the attendant was saying, I saw the woman ahead of me pull away with the hose still in her gas tank. In an instant, she had pulled the pump loose, and there was gasoline spraying in my face. At first it burned, and then I couldn't see. Luckily the attendants knew to flush out my eyes with lots of water. It took a while, but it worked, and my eyes are fine—a little red, but fine.

For some reason, I didn't panic when I was blinded. Everything else this week has managed to work out OK, so I guess I just trusted that this would, too. (Thank you, God!)

Of course, I missed my flight, and, it being Friday evening, it was hard to get another one. And because it was so dark when I did finally take off, I got to see Washington by night—a beautiful sight I would have missed had I flown out at 4:30 as planned. (Be in the moment, right?)

Gosh, it's good to be home.

### Monday, May 10

Things were really hectic at the office today, but because I'm teaching again tomorrow, I forced myself to make time to make the revisions to the course that I thought were most important;

they were mostly clarifications of workshop instructions. Also, I moved one exercise so it comes earlier in the course—we'll see how that works out. Then I cleaned up my foils and put them in order and photocopied my changes to the handouts. The workbooks had already been prepared, so luckily there were no changes I needed to make to them.

Barb and Henry and Lucy all asked me today how my trip was. I just smiled each time and answered, "Well, the plane didn't crash."

# *Quick Reference Checklists*

## *Preparation*

### *Prepare Yourself*

1. Know the subject.
2. Know the course.
3. Know your audience.
4. Visualize success.
5. Be well-rested.
6. Be set up and ready before participants arrive.

### *Prepare Your Presentation*

1. Have a walkthrough.
2. Practice your introduction.
3. Tape record your lecture.
4. Work through all exercises and workshops.

### *Prepare Your Materials*

1. References.
2. Teaching aids.
3. Participant aids.

## Prepare the Classroom

1. Room arrangement.
2. Instructor's table.
3. Computers/terminals.
4. Audio-visual equipment.
5. Course materials.

## Miscellaneous

1. Know class location.
2. Know class hours.
3. Know who to call if you have problems.

## Adult Learners

1. Are self-directing.
2. Are experienced.
3. Are ready to learn.
4. Are problem-oriented.

## Starting the Class

### Before Class

1. Be psychologically ready.
2. Be physically ready.
3. Know your audience.
4. Have classroom set up.
5. Be organized.
6. Be ready to start on time.
7. Know exactly how you will begin.

### In the Classroom

1. Create a friendly atmosphere.
2. Introduce yourself.
3. Have participants introduce themselves.

4. Give an overview.
5. Cover miscellaneous announcements.

### Dealing With Nervousness

*Before Class*

1. Practice with a tape recorder.
2. Practice in front of a mirror.
3. Have a walkthrough.
4. Anticipate questions.

## During Class

1. Take several deep breaths and let your body relax.
2. Bite your tongue so your mouth won't be dry.
3. Smile.
4. Laugh.
5. Establish your credibility during your introduction.
6. Use instructor notes.
7. Learn to read your audience.

### Controlling Class Flow

1. Be punctual.
2. Communicate control.
   • Speak with courteous authority.
   • Be assertive.
3. Know your course.
4. Handle interruptions confidently and courteously.

### Communication Skills

*Verbal*

1. Avoid hedge words ("I tend to think" "sort of" "I guess").
2. Avoid tag questions ("right?" "don't you think?").
3. Don't overdo intensifiers ("so" "really" "very"); diminutives ("little" "bitsy"); and rave words ("wow!" "super!" "awesome!").

## Nonverbal

1. Pitch—pleasantly low.
2. Volume—loud enough to be heard in class.
3. Speed—about 150 words per minute for teaching.
4. Voice quality—orally (not nasally) resonated.
5. Articulation—watch frequently used words and phrases.
6. Paralanguage—avoid "um" "er" "ah."

## Visual

1. Appearance.
2. Body language.
   • Posture
   • Movement
   • Gestures
   • Facial Expressions

3. Eye Contact.

## Conceptual

1. Who you are.
2. What you do.
3. Where you come from.
4. Your knowledge/experience.
5. Your status.

### Hands-on Workshops

1. Design workshops carefully.
2. Plan for workshops to immediately follow related lecture.
3. Give clear instructions.
4. Be available to assist participants.
5. Pay attention to feedback from participants during workshop.
6. End at the time you planned.
7. Discuss workshop afterward; make sure objectives were met.

## *Ending the Class*

1. Review.
   - major points
   - objectives

2. Are there questions?
3. What comes next?
4. How can you be reached?
5. Evaluations.

# Suggested Reading

## Articles

Selwyn Brent, "The American Trainer Abroad," *Data Training*, April 1988, p. 36.

Susan M. Bojar and Terrence Boothman, "Basic Training for Trainers," *Data Training*, January 1988, p. 34.

Phillip Bruschi, "Getting It By Ear—Listening in the Training Process," *Data Training*, April 1986, p. 29.

Robert M. Caldwell and Marvin Marcel, "Evaluating Trainers: In Search of the Perfect Method," *TRAINING*, January 1985, p. 52.

Ruth C. Clark and Phyllis Kyker, "How to Select Good Technical Instructors," *TRAINING*, December 1985, p. 55.

Lennie Copeland and Lewis Griggs, "On the Road Again," *Data Training*, April 1988, p. 43.

Chester Delaney, "Success + Success = Success Squared—A Training Imperative," *Data Training*, November 1985, p. 14.

Lucy D. Freedman and Lisa J. Marshall, "The Trainer as Communicator—Avoiding Common Errors," *Data Training*, April 1986, p. 26.

Jerry W. Gilley, "Adult Learners and the Classroom," *Data Training*, March 1987, p. 58.

Stephen Hughes, "Introductions and Icebreakers—Getting a Class Off the Ground," *Data Training*, October 1985, p. 16.

Stephen Hughes, "Ready For Anything," *Data Training*, November 1987, p. 74.

Stephen Hughes, "Ready, Willing, and Able," *Data Training*, January 1988, p. 14.

Stephen Hughes, "Who Said That?," *Data Training*, May 1988, p. 14.

Myron S. Kaplan, "Beating the Training-Stress Connection," *Data Training*, March 1988, p. 27.

Edward A. Kelly, "Are There Students in Your Classroom?," *Data Training*, March 1988, p. 21.

Lyle Lacmuith, "From Techie to Trainer," *Data Training*, January 1988, p. 18.

Antoni A. Louw, "Stage Fright (How to Break your Own Barriers and Become a More Effective Presenter)," *TRAINING*, August 1986, p. 55.

G.A. Marken, "Presentations With Punch," *MW*, March 1986, p. 14.

Bernard V. McGinley, "Still Doing Classroom Training? So Am I!," *Data Training*, September 1987, p. 32.

Robert F. Orgel and Ann I. Kelley, "Why Start at the Beginning? Non-Traditional Instructional Sequences," *Data Training*, October 1986, p. 42.

Sue Redkey, "Volunteer Instructors," *Data Training*, May 1988, p. 22.

Andrew E. Schwartz, "The Well-Tempered Classroom—Managing Disruptive Behavior," *Data Training*, January 1986, p. 30

Jane Stein, "DP Trainers Must Develop a High Tolerance for Heat," *Computerworld*, August 24, 1987, p. S16.

Linda Swirczek, "On the Road Again—Lives and Times of Traveling Instructors," *Data Training*, September 1985, p. 22.

Rebekah Wolman, "From Stress to Burnout," *Data Training*, November 1987, p. 47.

"Fifty Years of Training: An Interview with Malcolm Knowles," *Data Training*, January 1988, p. 19.

"Worst Moments in Training (As told to *TRAINING*'s editors)," *TRAINING*, September 1984, p. 27.

# Index

Adult learners
    characteristics, 3
    checklist for teaching, 8
    how they learn, 7
    how well they remember, 8
Affective (learning objectives), 54
Aids
    teaching, 30, 61
    types, 62
    visual, 8, 9, 22
Analogies, 80
Angled rows (room arrangement), 14
Appearance, 46
Articulation, 44
Asking questions, 88
Audience, knowing your general
        audience, 25
    how to find out what you need
        to know, 26
    specific audience, 26
    what you need to know, 25

Board (chalk), 63, 65
Body, care of
    emotional, 108
    mental, 107
    physical, 107

Body language, 46
Break-out rooms, 14

Cognitive (learning objectives), 53
Communication, effective, 37
Conceptual message, 48
Course, knowing your, 21, 76
Creative visualization, 31
Credibility, 71
Control, 71, 75
Controlling class flow, 74
Current, staying, 20, 22

Demonstrations, 57, 79, 101
Detail, 22
Diminutives, 40

Ending the class, 109
Evaluation approaches, 58
Examples, realistic, 6, 79
Exercises, 6, 22, 60, 79
Eye contact, 50

Facial expressions, 47
Feedback, 58, 88
Films, 63, 65
Flip charts, 62, 65
Flow of presentation, 22

Gestures, 47
Group tables (room arrangement), 13
Guided discussions, 9, 56, 79

Handouts, 64, 66
Hands-on workshops, 7, 9, 60, 80, 99
Hedge words, 39
Horseshoe (room arrangement), 12

"I don't know", 98
Illustrations, 80
Instructor notes, 22
Intensifiers, 40
Interruptions, 76
Introducing yourself, 72

Knowledge base, 108

Learning environment, 11, 71
Learning retention, 8
Lecture, 55
Listening, 49

Movement, 47, 104

Nervousness, 67
Networking, Appendix D, 125
Nonverbal message, 42 *See also*
    Conceptual message, Visual
    message, Vocal message

Oral reviews, 9
Overviews, 74

Paralanguage, 45
Physical comfort, 11, 107
Physical environment, 11
Pitch, 42
Posture, 46
Preparing
    the classroom, 30
    materials, 30
    the presentation, 29
    yourself, 29
Presentation methods, 53
Problem participants, 83

Professional presence, 109
Psychological atmosphere, 14
Psychomotor (learning objectives), 54

Rave words, 40

Questions
    asking, 88
    responding to answers, 91
    responding to questions, 94

Reviews, 60
Role play, 79
Room arrangement, 13

Scope of course, 21
Screen projection, 63
Slides, 64, 65
Speaking patterns, changing, 49
Speed, 43
Starting the class, 71
Subject, knowing your, 19, 80

Tag questions, 40
Team teaching, 81
Theater style (room arrangement), 12
Timing, 21
Traditional classroom style (room
    arrangement), 12
Transitions, 76
Transparencies, 62, 64

Verbal message, 39
Visual message, 45
Visualizing success, 31
Voice quality, 44
Volume (of speech), 43
Volunteer instructors, Appendix C,
    123
Vocal message, 42

Walkthroughs, 21, 33
    evaluations, Appendix A, 115
    Guidelines, 33
Workshops, hands-on, 7, 9, 60, 80,
    99